Springer Theses

Recognizing Outstanding Ph.D. Research

For further volumes:
http://www.springer.com/series/8790

Aims and Scope

The series "Springer Theses" brings together a selection of the very best Ph.D. theses from around the world and across the physical sciences. Nominated and endorsed by two recognized specialists, each published volume has been selected for its scientific excellence and the high impact of its contents for the pertinent field of research. For greater accessibility to non-specialists, the published versions include an extended introduction, as well as a foreword by the student's supervisor explaining the special relevance of the work for the field. As a whole, the series will provide a valuable resource both for newcomers to the research fields described, and for other scientists seeking detailed background information on special questions. Finally, it provides an accredited documentation of the valuable contributions made by today's younger generation of scientists.

Theses are accepted into the series by invited nomination only and must fulfill all of the following criteria

- They must be written in good English.
- The topic should fall within the confines of Chemistry, Physics and related interdisciplinary fields such as Materials, Nanoscience, Chemical Engineering, Complex Systems and Biophysics.
- The work reported in the thesis must represent a significant scientific advance.
- If the thesis includes previously published material, permission to reproduce this must be gained from the respective copyright holder.
- They must have been examined and passed during the 12 months prior to nomination.
- Each thesis should include a foreword by the supervisor outlining the significance of its content.
- The theses should have a clearly defined structure including an introduction accessible to scientists not expert in that particular field.

Christian Groß

Spin Squeezing and Non-linear Atom Interferometry with Bose–Einstein Condensates

Doctoral Thesis accepted by
the University of Heidelberg, Germany

Springer

Author
Dr. Christian Groß
University of Heidelberg
Heidelberg, Germany

Supervisor
Prof. Dr. Markus K. Oberthaler
University of Heidelberg
Heidelberg, Germany

ISSN 2190-5053
ISBN 978-3-642-43245-3
DOI 10.1007/978-3-642-25637-0
Springer Heidelberg Dordrecht London New York

e-ISSN 2190-5061
ISBN 978-3-642-25637-0 (eBook)

Springer is part of Springer Science+Business Media (www.springer.com)

Supervisor's Foreword

A thesis summarizes the essence, the final insight, the net-gain of knowledge of intensive research hours during 3 years. It is as such an invaluable contribution to the advance of science. Sometimes a doctoral thesis is regarded as an antique leftover. But it appears very modern and even an essential alternative for scientific exchange if it is seen in perspective to the nowadays typical form of scientific exchange via publications in highly reputed journals. There, the information is sometimes compactified to an extent that it becomes unreadable for non experts. Even for experts the streamlining of the text, optimized in order to meet the needs of the journals, makes some articles very inefficient for efficient information flow. The importance of the written thesis, where introductory as well as detailed information is conveyed, cannot be overestimated. The thesis following this foreword contains important additional information to the published results. It gives the reader an introduction to the concept of quantum metrology and spin squeezing as well as important information of what steps are crucial to realize this quantum resources in her/his own lab.

The theme of the research described in the thesis by Christian Gross falls into the category of Quantum Metrology. This concept is intimately associated with one of the very important paradigm shifts of the last century, where quantum mechanics was not any longer regarded as the loss of predictability, but more as a new approach for going beyond classical possibilities. Also in the more general framework of parameter estimation where classical statistical tools reveal the principle limits of precision we can nowadays utilizing quantum mechanical many particle states to overcome these limits of classical experiments—quantum Metrology. It was actually for the first time within Christian Gross' thesis that this concept of going beyond classical interferometry has been explicitly demonstrated for atomic matterwaves.

The thesis is suitable for experts as well as for newcomers in the field of quantum interferometry with matterwaves. A reader unfamiliar with the terminology of spin squeezing, its connection to entanglement and its application to interferometry will find a very valuable didactical introduction in the first chapter of the doctoral thesis. With this introduction the following two chapters going into the deep details of spin squeezing generation with Bose Einstein condensates, detection as well as its application to matterwave interferometry is readily accessible also for newcomers in the field and is enjoyable to read for the expert. At this point let me make a remark on the side: at the beginning of these experimental efforts it was not clear at all if the experimental stability can be pushed far enough that the quantum effects become observable and even useable. In both experimental campaigns described in detail in the third and forth chapter the final experimental breakthrough has been achieved with the last big change of the experimental setup after which the team would have changed the topic.

Concluding I would like to add a more general statement. Science is about asking questions, working hard on solving the problems and finally conveying the results efficiently to the community of scientists. Christian Gross' research activities during his thesis are a very good example where all three steps have been successfully taken and the reader will feel this spirit, while reading the thesis.

Heidelberg, October 2011   Prof. Dr. Markus K. Oberthaler

Acknowledgments

I want to thank my family, friends and colleagues who greatly supported me during my PhD studies.

- First of all I thank my wife Eva for all her love. She is always at my side "through thick and thin" and she took care for the necessary distraction from the lab work.
- In particular I want to thank my parents Monika und Michael for her loving support and the very relaxing weekends at home in the Westerwald.
- I thank my colleagues working at the NaLi, Argon, Atta and Aegis experiments for the cheerful atmosphere in our group and for the many hours on the mountain bike or at the climbing spots around Heidelberg.
- Thomas Gasenzer, Peter Schmelcher and Gershon Kurizki contributed to my thesis by many enlightening discussions.
- For the interesting discussions I thank all PhD students and post-docs who visited us at the Kirchhoff Institute or whom I met at conferences. In particular I want to thank Nir Bar-Gill, Dasari Durga Bhaktavatsala Rao and Giulia Ferrini.
- The electronics department at the KIP, in particular Jürgen Schölles, contributed strongly to our experiments. Jürgen provided us with very welcome relaxation during his yoga lessons.
- I thank the people from the mechanical workshop who helped setting up our experiments by complicated (the 'Eiffel-tower') as well as massive (40 kg aluminum block) constructions. I want to mention especially Mr. Spiegel, who helped to solve all the small (or big) problems with his straightforward help.
- All the people from the administration and computing department of the KIP provided help whenever it was needed.
- A very big "Thank You" goes to our team assistants Dagmar Hufnagel and Christiane Jäger for their helpful advise with any organizational needs. They contributed a lot to the cheerful atmosphere in the group.
- I also want to thank all the 'HiWis', who worked at our experiment and especially Rostislav Doganov, who assisted us with his sophisticated electronic circuits.

- The diploma students Timo Ottenstein, Jens Appmeier and Jens-Philipp Ronzheimer helped to put forward the experiment by several crucial technical contributions and ideas that enabled the presented experiments. Many thanks goes also to the new diploma students Helmut Strobel, Ion Stroescu and Wolfgang Müssel, who integrated into the team in a very short time. They are now crucially supporting the new experiments.
- Very important for the development of my understanding of our "double well" system were the detailed and basic discussions with Stefano Giovanazzi. Thank you Stefano.
- I want to thank my colleague Andreas Weller for the fruitful work together and for the thoughtful solutions of many experimental problems.
- Tilman Zibold, Eike Nicklas and myself made up a perfect team of PhD students at the BEC experiment. Thank you for the many constructive and lively discussions!
- I would like to give special thanks to Jérôme Estève for his indispensable support in developing the understanding for both theoretical and experimental issues of our work. I thank him not only for the joint work in the lab but also for the many private activities, e.g. sports climbing trips together with Bénédicte and Eva.
- Last but not least I want to thank my supervisor Markus Oberthaler. He is the perfect "boss", always open for ideas and discussions. Thank you Markus, for the big freedom when carrying out the experiments and for the open and fair leading of the group. Thank's for the support and giving me the opportunity to represent our group at many conferences—the discussions contributed a lot to my studies.

Finally I want to thank the Landesgratuiertenförderung Baden-Württemberg for the financial support during my PhD studies.

Contents

Chapter 1
Introduction

Today's most precise measurement instruments work at the shot noise limit, the precision bound set by single particle quantum mechanics. Many of these devices are interferometers, based on the interference of two atomic or photonic quantum states. The observable to be measured causes a relative phase shift φ between the two modes of the interferometer. This relative phase shift is observed indirectly as a population imbalance at readout. In the readout process the population imbalance is obtained by counting the atoms or photons in each of the modes and their particle-like properties become important.

For uncorrelated quantum states single particle quantum mechanics describes the measurement process. The relative phase φ determines the probability for each atom or photon to be detected in one of the modes—in the balanced case $\varphi = 0$ the probability for both modes is equal. The probability distribution of the population imbalance is poissonian and the measurement uncertainty in the relative phase $\Delta\varphi^2$ scales statistically as $\Delta\varphi^2 = 1/N$. Thus, the shot noise limit for the measurement precision arises as the classical statistical limit of N uncorrelated particles used in the interferometer [1, 2]. One single measurement with N independent resources is equivalent to N identical measurements using only one resource. Fundamentally this "classical" noise results from the projection of the quantum state on the two observed output states in the readout process. Equivalent to the term *shot noise* commonly *quantum projection noise* is used and the resulting precision limit is the *standard quantum limit*.

Photonic interferometers are commonly used for distance or velocity sensors [3] while prominent examples for atomic sensors are measurements of magnetic fields, inertia or time [4].

Many-body quantum mechanics offers the possibility to overcome the single particle limit by the use of entanglement as a resource. Focussing on atom interferometry, different quantum strategies have been proposed to obtain interferometric precision beyond "classical" bounds [5–8]. The quantum Cramer-Rao bound reveals the fundamental limit [2], the so called *Heisenberg limit for metrology*, where the obtainable phase precision scales as $\Delta\varphi^2 = 1/N^2$. The potential gain is enormous. In an atomic

C. Groß, *Spin Squeezing and Non-linear Atom Interferometry with Bose–Einstein Condensates*, Springer Theses, DOI: 10.1007/978-3-642-25637-0_1,
© Springer-Verlag Berlin Heidelberg 2012

clock for example one measures an energy difference $\hbar\omega$ for a certain time t ($2\pi\hbar$ is Planck's constant). The "classical" uncertainty scales as $\Delta\omega^2 = 1/tN$, while Heisenberg limited metrology would allow for an error of $\Delta\omega^2 = 1/tN^2$. Assuming a fictitious Heisenberg limited measurement with 10^6 atoms lasting one second, a "classical" projection noise limited apparatus would need 11 days to obtain the same level of precision.

Spin squeezing is one example where entanglement provides a resource for quantum enhanced metrology. For atomic two-level systems the concept of spin squeezed states and a mechanism to obtain them was introduced by Kitagawa and Ueda in 1993 [6]. One year later Wineland et al. pointed out its potential usefulness for atom interferometry [5]. The basic idea is to use a quantum correlated spin state for Ramsey type interferometry [9, 10], where the quantum fluctuations in the different spin directions are redistributed. Atomic spin squeezing has already been experimentally demonstrated in vapor cell experiments [11–14], ion traps [15] and recently with laser cooled atoms [16,17]. We demonstrate spin squeezing in a Bose–Einstein condensate where distinct to vapor cell experiments the center of mass motion of the atoms is controlled and where many particles contribute to a single measurement, which is a limitation in ion trap experiments.

In the following paragraph we briefly outline the connection of spin squeezing to interferometric phase estimation precision. Any two-mode system, and therefore any quantum state in a two-mode interferometer, can be described by a fictitious spin vector with total length J. In a symmetric situation, valid for N Bosons in two modes, the Schwinger representation [18] connects the three orthogonal components of the spin vector to the creation and annihilation operators $\hat{a}(\hat{b})$ and $\hat{a}^\dagger(\hat{b}^\dagger)$ of the two modes. A direct relation between occupation number difference n and relative phase φ on one side and the spin operators ($\hat{J}x$, $\hat{J}y$, $\hat{J}z$) on the other side exists [19]. Therefore engineering of the quantum fluctuations in the different spin directions can be used to obtain a quantum state that features reduced fluctuations in the relative phase φ, the quantity of interest in interferometry. This is the basic mechanism to obtain quantum enhanced precision beyond the shot noise limit with spin squeezed states. However, nature forbids to reduce the variance in the relative phase φ arbitrarily since the population difference n is its conjugate variable. According to Heisenberg's uncertainty principle a decrease of the phase variance causes an increase of the population difference fluctuations which eventually degrades interferometric precision. Therefore knowledge of the fluctuations in both conjugate variables is important to characterize the usefulness of a quantum state for interferometry.

An uncorrelated collective spin state with mean spin pointing for example in J_x direction has isotropic fluctuations in the orthogonal spin directions J_z and J_y. Redistribution of these fluctuations requires quantum correlations between the different constituents. Therefore enhanced interferometric sensitivity in atom interferometers is connected to entanglement among the atoms. In 2001 Sørensen et. clarified the connection between metrologically relevant spin squeezing and entanglement [20, 21].

In this thesis we report on experiments detecting many-body entanglement in a Bose–Einstein condensate of ^{87} Rubidium atoms. Two different experimental systems are used. We achieve coherent spin squeezing among two external degrees of freedom of the condensate—two mean field modes—populated with approximately 2,000 atoms. Despite of finite temperature in the system we observe up to $\xi_S^2 = -3.8$ dB coherent spin squeezing where ξ_S^2 is the parameter that quantifies the potential amount of precision gain in interferometry. Spin squeezed states are engineered employing an adiabatic cooling approach where temperature induced fluctuations are reduced such that spin squeezing grows. Finite initial temperature and therefore higher entropy in the system is identified as the limiting factor.

In a second set of experiments we demonstrate coherent spin squeezing between two internal degrees of freedom of the condensate—two hyperfine states. Microwave and radio frequency coupling pulses allow for a very accurate control of the collective spin vector and a narrow magnetic Feshbach resonance is used to tune the interatomic interactions. We realize a novel non-linear atom interferometer and measurements on 400 atoms directly demonstrate 15% enhanced interferometric precision beyond the standard quantum limit. Characterization of the quantum state within the interferometer reveals $\xi_S^2 = -8.2$ dB coherent spin squeezing. This requires the presence of 170 entangled particles [20] and we exclude less than 80 entangled particles with three standard deviation statistical confidence. These experiments are done at zero effective temperature, but loss of the atoms from the trap is identified as the limit for the obtainable coherent spin squeezing.

This thesis is organized as described below. After Chap. 1, this introduction, we review the basic concepts of spin squeezing and its connection to many-body entanglement and interferometry in Chap. 2. The following Chap. 3 deals with the experiments done with a single component Bose–Einstein condensate in external double- and few-well potentials. The results of these experiments have been published in reference [22]. In the last Chap. 4 we report on the realization of a non-linear atom interferometer and we directly show measurement precision beyond the standard quantum limit. Our findings have been published in [23]. A comprehensive appendix on precision absorption imaging with high spatial resolution describes the detection method used for the experiments. Throughout the thesis we use the most intuitive units for the energy E. Either angular frequency $\omega = E/\hbar$ or temperature $T = E/k_B$ is given where k_B is Bolzmann's constant. $\hbar$ and k_B are normalized to unity and it is useful to remember the conversion between angular frequency and temperature $\omega/T \approx 2\pi \times 20$ Hz/nK.

Not directly related to this theses but measured at the same time we published a paper on *Experimental observation of oscillating and interacting matter wave dark solitons* [24].

References

1. Giovannetti V, Lloyd S, Maccone L (2004) Quantum-enhanced measurements: beating stand quantum limit. Science 306:1330–1336
2. Giovannetti V, Lloyd S, Maccone L (2006) Quantum metrology. Phys Rev Lett 96:010401
3. Hariharan P (2003) Optical interferometry. Academic, London

4. Cronin AD, Schmiedmayer J, Pritchard DE (2009) Optics and interferometry with atoms and molecules. Rev Mod Phys 81:1051
5. Wineland D, Bollinger J, Itano W, Heinzen D (1994) Squeezed atomic states and projection noise in spectroscopy. Phys Rev A 50:67–88
6. Kitagawa M, Ueda M (1993) Squeezed spin states. Phys Rev A 47:5138–5143
7. Bouyer P, Kasevich MA (1997) Heisenberg-limited spectroscopy with degenerate Bose–Einstein gases. Phys Rev A 56:1083–1086
8. Dowling JP (1998) Correlated input-port, matter-wave interferometer: quantum-noise limits to the atom-laser gyroscope. Phys Rev A 57:4736–4746
9. Ramsey NF (1949) A new molecular beam resonance method. Phys Rev 76:996
10. Ramsey NF (1950) A molecular beam resonance method with separated oscillating fields. Phys Rev 78:695–699
11. Hald J, Sørensen JL, Schori C, Polzik ES (1999) Spin squeezed atoms: a macroscopic entangled ensemble created by light. Phys Rev Lett 83:1319–1322
12. Kuzmich A, Mandel L, Bigelow NP (2000) Generation of spin squeezing via continuous quantum nondemolition measurement. Phys Rev Lett 85:1594–1597
13. Fernholz T et al (2008) Spin squeezing of atomic ensembles via nuclear-electronic spin entanglement. Phys Rev Lett 101:073601
14. Appel J et al (2009) Mesoscopic atomic entanglement for precision measurements beyond the standard quantum limit. Proc Natl Acad Sci USA 106:10960–10965
15. Meyer V et al (2001) Experimental demonstration of entanglement-enhanced rotation angle estimation using trapped ions. Phys Rev Lett 86:5870–5873
16. Schleier-Smith MH, Leroux ID, Vuletic V (2010) Reduced-quantum-uncertainty states of an ensemble of two-level atoms. Phys Rev Lett 104:73604
17. Leroux ID, Schleier-Smith MH, Vuletic V (2010) Implementation of cavity squeezing of a collective atomic spin. Phys Rev Lett 104:73602
18. Sakurai J (1994) Modern quantum mechanics. Addison-Wesley, Reading
19. Leggett A (2001) Bose–Einstein condensation in the alkali gases: some fundamental concepts. Rev Mod Phys 73:307–356
20. Sørensen AS, Mølmer K (2001) Entanglement and extreme spin squeezing. Phys Rev Lett 86:4431–4434
21. Sørensen AS, Duan L, Cirac J, Zoller P (2001) Many-particle entanglement with Bose–Einstein condensates. Nature 409:63–6
22. Estéve J, Gross C, Weller A, Giovanazzi S, Oberthaler MK (2008) Squeezing and entanglement in a Bose–Einstein condensate. Nature 455:1216–1219
23. Gross C, Zibold T, Nicklas E, Estéve J, Oberthaler MK (2010) Nonlinear atom interferometer surpasses classical precision limit. Nature 464:1165–1169
24. Weller A et al (2008) Experimental observation of oscillating and interacting matter wave dark solitons. Phys Rev Lett 101:130401

Chapter 2
Spin Squeezing, Entanglement and Quantum Metrology

Spin squeezing is a quantum strategy introduced in 1993 by Kitagawa and Ueda [1] which aims to redistribute the fluctuations of two conjugate spin directions among each other. In 1994 it was theoretically shown that spin squeezed states are useful quantum resources to enhance the precision of atom interferometers [2] and in 2001 the connection between spin squeezing and entanglement was pointed out [3].

In this chapter we introduce the spin representation for N two-level atoms. We review the basic theoretical concepts of spin squeezing and its connection to entanglement. Different entanglement criteria are discussed and the usefulness of entanglement as a resource in quantum metrology—focussing on spin squeezed states—is reviewed.

2.1 Collective Spins

The mathematical concept of a spin algebra with total spin J is a powerful tool to describe very different physical systems. Any observable within a spin J system can be expressed by the three spin operators $\hat{J}_x$, $\hat{J}_y$, $\hat{J}_z$ and the identity operator. The $2J + 1$ eigenstates of one of the spin operators make up a basis set of the $2J + 1$ dimensional Hilbert space. The choice of the direction is arbitrary since the operators are connected via unitary transformations.

2.1.1 A Single Spin 1/2 on the Bloch Sphere

One of the simplest nontrivial models in quantum mechanics, a two-level system [4] with levels $|a\rangle$ and $|b\rangle$, maps onto a spin $J = 1/2$ system. This mapping is done by assigning the state $|a\rangle$ to the eigenstate of $\hat{J}_z$ with eigenvalue $j_z = -1/2$ (spin down) and state $|b\rangle$ to the eigenstate with eigenvalue $j_z = +1/2$ (spin up). Two important applications of this model in atomic physics are the two-level atom and nuclear magnetic resonance experiments. Any pure quantum state

C. Groß, *Spin Squeezing and Non-linear Atom Interferometry with Bose–Einstein Condensates*, Springer Theses, DOI: 10.1007/978-3-642-25637-0_2,
© Springer-Verlag Berlin Heidelberg 2012

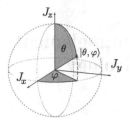

Fig. 2.1 The Bloch sphere. Schematic representation of the quantum state $|\theta, \varphi\rangle$ of a spin 1/2 system on the Bloch sphere. The definition of the longitudinal angle φ and the polar angle θ are highlighted and in the following the same notation will be used for the direction of the collective spin on a generalized Bloch sphere (see Sect. 2.2.2)

$|\theta, \varphi\rangle = \sin(\theta/2) |a\rangle + \cos(\theta/2) e^{i\varphi} |b\rangle$ of a two-level system can be conveniently represented on a *Bloch sphere*. The coordinate axes are chosen such that the population difference $(|b\rangle\langle b| - |a\rangle\langle a|)/2$ maps to the $\hat{J}_z$ component of the spin and the coherences $(|b\rangle\langle a| + |a\rangle\langle b|)/2$ and $(|b\rangle\langle a| - |a\rangle\langle b|)/2i$ map to the $\hat{J}_x$ and $\hat{J}_y$ components respectively. Figure 2.1 shows the quantum state on the Bloch sphere with the definition of the longitudinal angle φ and the polar angle θ. The Hilbert space for a single spin 1/2 system is two dimensional, such that the representation on the surface of the Bloch sphere does not require any additional assumptions.

2.1.2 A Large Collective Spin

The discussion above can be generalized for N particle systems where each particle is restricted to two modes. Each particle is an elementary spin $j = 1/2$ system, sometimes called *Qubit*.

The collective spin operators $\hat{J}_i$ can be defined as the sum over all elementary spin operators (Pauli matrices) $\hat{\sigma}_i^{(k)}$, where $i = (x, y, z)$:

$$\hat{J}_i = \sum_{k=1}^{N} \hat{\sigma}_i^{(k)} \tag{2.1}$$

A basis of the general problem can be obtained as the tensor product of all N bases of the individual components, each of dimension $(2j + 1)$. The dimension of the Hilbert space is huge $\dim(\mathcal{H}_N) = (2j + 1)^N = 2^N$ and grows exponentially with the number of Qubits. The length of the collective spin $\mathcal{J}$ is smaller or equal than half the number of Qubits[1]:

$$\sqrt{\mathcal{J}(\mathcal{J} + 1)} = \langle \hat{J}^2 \rangle^{1/2} \leq N/2 \tag{2.2}$$

One often assumed simplification is exchange symmetry among all Qubits. This is physically motivated since in many experiments all operations done on the ensemble

---

[1] In this thesis we deal with large spins such that we often approximate $\sqrt{\mathcal{J}(\mathcal{J} + 1)} \approx \mathcal{J}$.

affect each spin in the same way. One example are nuclear magnetic resonance experiments in homogeneous fields.

In the symmetric case each elementary Qubit can be prepared for example in the $j_z = -1/2$ state and maximum collective polarization $J_z = -N/2$ can always be reached. Therefore the total spin length is given by $J = N/2$ and the dimension of the Hilbert space dramatically reduces to $\dim(\mathcal{H}_S) = (2Nj + 1) = (N + 1)$, linearly dependent on the number of Qubits. One possible choice of a basis are the symmetric *Dicke states* $|J, m\rangle$ with $-N/2 < m < N/2$. Due to their exchange symmetry the elementary spins can be effectively described as Bosons, the *Schwinger Bosons* [5]. Employing the second quantization formalism the creation and annihilation operators of the two modes $\hat{a}^\dagger$ ($\hat{b}^\dagger$) and $\hat{a}$ ($\hat{b}$) can be related to the different spin components [6]:

$$\hat{J}_+ = \hat{b}^\dagger \hat{a}$$
$$\hat{J}_- = \hat{a}^\dagger \hat{b}$$
$$\hat{J}_x = \frac{1}{2}(\hat{J}_+ + \hat{J}_-)$$
$$\hat{J}_y = \frac{1}{2i}(\hat{J}_+ - \hat{J}_-)$$
$$\hat{J}_z = \frac{1}{2}(\hat{b}^\dagger \hat{b} - \hat{a}^\dagger \hat{a})$$

Each of the Dicke states introduced above corresponds to a perfectly defined particle number difference between the two modes $\hat{a}$ and $\hat{b}$ and since the total number of particles N is fixed the Dicke states correspond to *Fock states* in the two modes $\hat{a}$ and $\hat{b}$.

The experiments presented in this thesis deal with two-mode Bose–Einstein condensates. Identical particles in two modes (as the Bosons in the condensate) can be described by the symmetric spin model and the *Schwinger representation* given above is used to relate the creation and annihilation operators of the two modes to the different spin components.

Even if not formally correct we will use the notation J instead of $\mathcal{J}$ for all spins regardless of symmetry and mention explicitly where the symmetry argument is necessary.

2.2 Fluctuation Engineering

The three different orthogonal spin components are conjugate variables. Their commutation relation is $[\hat{J}_i, \hat{J}_j] = i\epsilon_{ijk}\hat{J}_k$, where ϵ_{ijk} is the Levi-Civita symbol. Therefore any pair of spin operators obeys a Heisenberg uncertainty relation which—for $\Delta \hat{J}_z^2$ and $\Delta \hat{J}_y^2$—is given by

$$\Delta \hat{J}_z^2 \Delta \hat{J}_y^2 \geq \frac{1}{4} \langle \hat{J}_x^2 \rangle \tag{2.3}$$

and $\Delta \hat{J}_z^2 = \langle \hat{J}_z^2 \rangle - \langle \hat{J}_z \rangle^2$ is the variance in $\hat{J}_z$ direction.

2.2.1 Coherent Spin States

Coherent spin states are the most classical-like pure quantum states of a symmetric ensemble of N elementary $j = 1/2$ spins or of N two-mode Bosons [6, 7]. They are constructed by placing all N particles in the same single particle state in any superposition of the two modes

$$|\theta, \varphi\rangle = \frac{1}{\sqrt{N!}} [\sin(\theta/2)\hat{a}^\dagger + \cos(\theta/2)e^{i\varphi}\hat{b}^\dagger]^N |\text{vac}\rangle \tag{2.4}$$

where $|\text{vac}\rangle$ is the vacuum state. Especially no quantum correlations between the particles are present. Therefore a coherent spin state features equal variance in any direction $\hat{J}_\perp$ orthogonal to the mean spin direction (θ, φ) which is given by the sum of the variances of the $2J$ elementary spin 1/2 particles. The perpendicular variances $\Delta \hat{\sigma}_\perp^2$ of individual Qubits are by definition isotropic around (θ, φ) since there are no subsystems that could cause any correlations [1]. The Heisenberg limit (2.2) for a single elementary spin pointing in σ_x direction is $\Delta \sigma_z^2 \Delta \sigma_y^2 = \frac{1}{4} \cdot \frac{1}{4}$ leading to an isotropic variance of

$$\Delta \hat{J}_z^2 = \Delta \hat{J}_y^2 = 2J \cdot \frac{1}{4} = \frac{J}{2} \tag{2.5}$$

for the collective coherent spin state, which identifies this quantum state as a minimal uncertainty state since $\langle \hat{J}_x \rangle = J$. We refer to the perpendicular spin fluctuations of a coherent spin state $\Delta \hat{J}_\perp^2 = J/2 = N/4$ as the *shot noise limit*.

We go back to the first quantization formalism in order to obtain the probability distribution over different sets of basis states—especially the two possible Dicke state bases in the directions orthogonal to the mean spin direction. In order to develop a more detailed understanding of the coherent spin state and its fluctuations we start with the discussion of a special case where each particle is in a 50/50 superposition of the two modes with 0 relative phase—each spin points in σ_x direction and its quantum state is

$$|x\rangle = \left(\left| \frac{1}{2}, -\frac{1}{2} \right\rangle + \left| \frac{1}{2}, +\frac{1}{2} \right\rangle \right) / \sqrt{2} \tag{2.6}$$

where we have chosen the Dicke states in σ_z direction as the basis states. The probability to observe each individual elementary spin in state up or down is equal

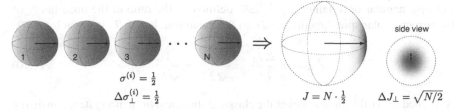

Fig. 2.2 A coherent spin state composed of elementary spins. The figure illustrates the addition of N elementary Qubits with equal mean spin orientation (*indicated by the arrows*) to a large collective spin J. The gray shading on the Bloch spheres visualizes the spread of the quantum state on the sphere using the Q-representation introduced in Sect. 2.2.2. The isotropic angular uncertainty decreases with the number of Qubits according to the standard quantum limit

$\left| \langle \frac{1}{2}, \pm \frac{1}{2} | x \rangle \right|^2 = 1/2$. The N atom coherent spin state is a collection of these independent elementary spins

$$|X\rangle = \left[\left(\left| \frac{1}{2}, -\frac{1}{2} \right\rangle + \left| \frac{1}{2}, +\frac{1}{2} \right\rangle \right) / \sqrt{2} \right]^{\otimes N} \tag{2.7}$$

and therefore the measurement of the J_z spin component is equivalent to N measurements on a single spin and the probability distribution over the Dicke states is binomial. We could have chosen equally the Dicke states in J_y direction to describe the spin state which shows again that the spin fluctuations in the directions perpendicular to J_x—the mean spin direction—are isotropic.

A general coherent spin state $|\theta, \varphi\rangle$ described as superposition of Dicke states $|J, m\rangle$ is given by [8]:

$$|\theta, \varphi\rangle = \sum_{m=-J}^{J} c_m(\theta) e^{-i(J+m)\varphi} |J, m\rangle \tag{2.8}$$

As argued above the coefficients $c_m(\theta)$ follow a binomial distribution peaked around θ:

$$c_m(\theta) = \binom{2J}{J+m}^{1/2} \cos(\theta/2)^{J-m} \sin(\theta/2)^{J+m} \tag{2.9}$$

Figure 2.2 depicts the composition of a large collective spin from elementary spins on generalized Bloch spheres.[2] The illustration of the spins is done using the Q-representation described in Sect. 2.2.2.

The Standard Quantum Limit

Due to the Heisenberg uncertainty principle (2.2) the mean direction (θ, φ) of any spin state can not be defined with arbitrary precision. For a coherent spin state the

---

[2] Above we give an example for the mean spin in J_x direction, however for the purpose of better illustration we have chosen a different direction in the figure.

isotropic angular uncertainty $\Delta\varphi = \Delta\theta$, defined by the ratio of the uncertainty of the perpendicular spin directions $\Delta J_\perp$ to the mean spin length J, is given by:

$$\Delta\varphi = \frac{\Delta\hat{J}_\perp}{\langle\hat{J}\rangle} = \frac{1}{\sqrt{2J}} = \frac{1}{\sqrt{N}} \tag{2.10}$$

As argued above this limit arises as the classical statistical limit in a system consisting of N independent constituents [2, 9]. In Sect. 2.4 we discuss the connection of spin states and Ramsey interferometry and we show that the angular uncertainty limits the interferometric precision. In this context the "classical" limit (2.10) for a coherent spin state is known as the *standard quantum limit*. Figure 2.2 also visualizes the decreasing angular uncertainty with the number of elementary spins.

2.2.2 Visualizing Spin States: The Husimi Q-Representation

Employing the Q-representation [10], spin states can be conveniently visualized on a generalized Bloch sphere with radius J. In order to describe the most general spin state, i.e. pure states and statistical mixtures, the density matrix formalism is used [6]. The density operator $\hat{\rho}$ in coherent spin state basis is given by

$$\hat{\rho} = \int P(\theta, \varphi)|\theta, \varphi\rangle\langle\theta, \varphi|d\Omega \tag{2.11}$$

where the integral covers the full solid angle and $d\Omega = \sin(\theta)d\theta d\phi$. The probability distribution $P(\theta, \varphi)$ is normalized to one. The Q-representation uses the diagonal elements of the density operator to represent the quantum state:

$$Q(\theta, \varphi) = \frac{2J + 1}{4\pi}\langle\theta, \varphi|\rho|\theta, \varphi\rangle \tag{2.12}$$

The interpretation of this representation on generalized Bloch spheres differs from the single spin $j = 1/2$ Bloch sphere shown in Fig. 2.1. In the latter case the dimension of the Hilbert space is two-dimensional and the quantum state representation on the surface of a sphere is exact. However for collective spin systems the dimension of the Hilbert space is $2J + 1$ such that an exact mapping to the surface of a sphere is not possible. The position (θ, ϕ) on the spin 1/2 Bloch sphere describes the full quantum state, while the position on the generalized Bloch sphere gives only the mean spin direction and—within the constraints explained below—its fluctuations. The Q-representation projects the density matrix on minimal uncertainty states, in particular coherent spin states. The most obvious consequence is that the minimal extension of a quantum state in (θ, φ) on the Bloch sphere is given by the uncertainties of the basis states—a single Dicke state features no uncertainty in polar direction but its Q-representation shows $\Delta\theta \propto 1/\sqrt{N}$.

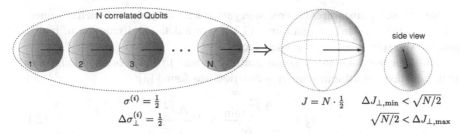

$$\sigma^{(i)} = \tfrac{1}{2}$$
$$\Delta\sigma_\perp^{(i)} = \tfrac{1}{2}$$

$$J = N \cdot \tfrac{1}{2} \qquad \Delta J_{\perp,\mathrm{min}} < \sqrt{N/2}$$
$$\sqrt{N/2} < \Delta J_{\perp,\mathrm{max}}$$

Fig. 2.3 Spin squeezed states. The figure illustrates an exemplary pure spin squeezed state on the Bloch sphere. The individual Qubits feature an isotropic variance, but quantum correlations between them cause an anisotropic variance of the collective spin state. For a Heisenberg limited spin squeezed state, one of the perpendicular variances $\Delta J_{\perp,\mathrm{min}}$ is smaller than the variance of a coherent spin state (of the same spin length), while the variance in the second perpendicular direction $\Delta J_{\perp,\mathrm{max}}$ is increased

2.2.3 Spin Squeezed States

Quantum correlations between the elementary spin 1/2 particles of a collective spin J can cause anisotropic fluctuations of the spin vector in the directions perpendicular to the mean spin (Fig. 2.3). Nevertheless the fluctuations of each individual elementary spin are always isotropic [1]. In Ref. [1] quantum states are considered spin squeezed if the variance of one spin component is smaller than the shot noise limit $J/2$ for a coherent spin state:

$$\xi_N^2 = \frac{2}{J}\Delta\hat{J}_{\perp,\mathrm{min}}^2 \tag{2.13}$$

Definition (2.13) does not take the second perpendicular spin direction into account. Due to the Heisenberg uncertainty relation (2.2) reduction of the variance in one direction causes an increase of fluctuations in the other. Real life strategies to obtain spin squeezing might also involve states that are not minimal uncertainty states. One example is the "one axis twisting" scheme proposed in [1], which we use in the experiments described in the last chapter of this thesis. For these states, as for experimentally very important non-pure quantum states, the variance in some other direction than the squeezed direction can be much larger than given by the Heisenberg uncertainty relation. This leads to a reduction of the effective mean spin length $\langle \hat{\mathbf{J}} \rangle$.

Metrologic applications, especially Ramsey interferometry for which spin squeezed states have been considered useful, require a large mean spin length. In order to measure the usefulness of spin squeezed states for these applications another definition of the squeezing parameter was introduced in Ref. [2]

$$\xi_R = \sqrt{2J}\frac{\Delta\hat{J}_{\perp,\mathrm{min}}}{\langle \hat{\mathbf{J}} \rangle} \tag{2.14}$$

whose inverse ξ_R^{-1} measures the precision gain in a Ramsey interferometric sequence relative to the standard quantum limit (2.10). For a detailed discussion of interferometry with spin squeezed states see Sect. 2.4

Spin squeezing among N constituents is related to *many-body entanglement*. In this context a third spin squeezing criterion was found [3]:

$$\xi_S^2 = N \frac{\Delta \hat{J}_{\perp,\min}^2}{\langle \hat{\mathbf{J}} \rangle^2} = N \frac{\Delta \hat{J}_z^2}{\langle \hat{J}_x \rangle^2} \tag{2.15}$$

Entanglement is detected by the inequality $\xi_S^2 < 1$ as detailed in the following section. Here we explicitly use the standard assumption throughout this thesis that the mean spin points in J_x direction and the direction of minimal variance—if not explicitly mentioned—is the J_z direction.

ξ_S can be used equivalently to ξ_R to quantify spin squeezing and precision gain in interferometry and we refer to it as *coherent number squeezing* or *coherent spin squeezing*.

2.3 Spin Squeezing and Entanglement

2.3.1 Definition of Many-Body Entanglement

For N distinguishable particles the definition of a separable state, i.e. non-entangled state, is that its N-body density matrix ρ can be written as a direct product of single particle density matrices $\rho^{(i)}$:

$$\rho = \sum_k p_k \rho_k^{(1)} \otimes \rho_k^{(2)} \otimes \cdots \otimes \rho_k^{(N)} \tag{2.16}$$

p_k is a probability distribution to account for incoherent mixtures. Entanglement in many-body systems (for a general review see [11, 12]) is defined as the non-separability of the density matrix ρ meaning the equality in Eq. 2.16 does not hold.

In collective spin systems a separable state is composed of independent elementary spin 1/2 particles. Due to technical limitations the individual elementary spins can not be addressed in many experiments . However it is important to note that the elementary spins have to be in principle distinguishable in order to define entanglement among them in a meaningful way [11]. In the scope of this thesis we deal with N particles in a Bose–Einstein condensate where the distinguishability is not obvious. However Sørensen and Mølmer pointed out that by a gedanken local operation one can pinpoint each particle in space without affecting the spin properties of the system (Sørensen AS, Mølmer K private communication). The distinguishability is now given via the position of each particle. If entanglement is detected in the system, it must have been present in the system before the localization, since local measurements can not generate entanglement [13]. Given that the atoms in the Bose–Einstein condensate are spaced by more than one wavelength of the detection light (which is usually fulfilled), this gedanken local operation means to overcome the technical limitations for addressability and detection of the individual Qubits.

The question of entanglement in bosonic pseudo spin systems and its connection to spin squeezing are extensively discussed in [14].

2.3.2 Entanglement Criteria Based on Collective Spin Variables

Without the possibility to address the individual Qubits entanglement criteria based on the collective spin variables are necessary to detect entanglement. Furthermore the observables in most experiments so far are limited to first and second order moments of the distributions functions in different spin directions due to rather small counting statistics and technical noise. Based on these, a complete set of inequalities that is fulfilled for any separable quantum state has been found [15, 16]. Complete in this sense means that assuming the only information available are first ($\langle \hat{J}_x \rangle$, $\langle \hat{J}_y \rangle$, $\langle \hat{J}_z \rangle$) and second moments ($\Delta \hat{J}_x^2$, $\Delta \hat{J}_y^2$, $\Delta \hat{J}_z^2$) of the distribution functions. These inequalities are:

$$\langle \hat{J}_x^2 \rangle + \langle \hat{J}_y^2 \rangle + \langle \hat{J}_z^2 \rangle \leq \frac{N(N+2)}{4} \tag{2.17}$$

$$\Delta \hat{J}_x^2 + \Delta \hat{J}_y^2 + \Delta \hat{J}_z^2 \geq \frac{N}{2} \tag{2.18}$$

$$\langle \hat{J}_i^2 \rangle + \langle \hat{J}_j^2 \rangle - \frac{N}{2} \leq (N-1)\Delta \hat{J}_k^2 \tag{2.19}$$

$$(N-1)[\Delta \hat{J}_i^2 + \Delta \hat{J}_j^2] \geq \langle \hat{J}_k^2 \rangle + \frac{N(N-2)}{4} \tag{2.20}$$

Toth et al. published these inequalities in Refs. [15, 16] and the authors depict the inequalities by a volume containing all separable states in a three dimensional space spanned by ($\Delta \hat{J}_x^2$, $\Delta \hat{J}_y^2$, $\Delta \hat{J}_z^2$).

Throughout this thesis we use the original spin squeezing inequality (2.15) in order to detect spin squeezing type entanglement experimentally [3]. All separable states fulfill the inequality $\xi_S^2 \geq 1$, but a subgroup of entangled states violate it.

As pointed out in Ref. [16], this criterion is equivalent to criterion (2.20) in the limit of large N and the mean spin pointing in J_x direction.

None of the entanglement witnesses given in this section requires any symmetry assumption. They are valid for the general definition of the collective spin given in Eq. 2.1. Entanglement criteria only valid under the symmetric two-mode assumption are discussed in the next section.

Entanglement Criteria for Symmetric States

Making the strong assumption of symmetry under particle exchange many entanglement criteria simplify. In this case the detection of spin fluctuations in one direction below the shot noise limit for N atoms implies entanglement [17–20].

$$\xi_N^2 = \frac{4\Delta \hat{J}_{\perp,\mathrm{min}}^2}{N} = \frac{2\Delta \hat{J}_{\perp,\mathrm{min}}^2}{J} \geq 1 \tag{2.21}$$

holds for any separable symmetric state. For clarity the mean spin is assumed to point in J_x direction such that $\langle \hat{J}_{\perp,\mathrm{min}} \rangle = 0$.[3] Equation (2.21) is identical to the spin squeezing definition of Kitagawa and Ueda (2.13) showing that at least in the symmetric two-mode case entanglement is necessary to redistribute the fluctuations of orthogonal spin components. Within this thesis we refer to ξ_N^2 as *number squeezing*.

All entanglement witnesses discussed here are based on second moments, therefore they contain maximally two body correlations $\langle \hat{\sigma}_k^{(i)} \hat{\sigma}_k^{(j)} \rangle$ of the elementary spins i and J in direction k. The question arises if these criteria detect only *bipartite entanglement*, the non-separability of the average two-body density matrix.

Toth, et al. show in Ref. [16] that in the non-symmetric case the complete set of separability criteria (Eqs. 2.17–2.20) can detect entanglement even if there is no bipartite entanglement in the system—the average two-body density matrix of an nonsymmetric state can be separable even if the N-body density matrix is entangled. The situation is different in the symmetric case. Here the violation of the number squeezing criterion (2.21) is both necessary and sufficient for bipartite entanglement in the system. Every bipartite entangled symmetric state features number squeezing [17].

2.3.3 Experimentally Used Quantification of Entanglement

The criteria given above are useful to detect the presence of entanglement, however they do not quantify entanglement in the system.[4] Two experimentally used approaches to quantify entanglement are reviewed here.

Von Neumann Entropy

In a recent experiment entanglement has been reported based on the *von Neumann entropy* [21]. However, we clarify in this short section that it is not possible to characterize entanglement in our experimental system by this measure.

For *pure* quantum states the von Neumann entropy $S_N(\hat{\rho}_A) = -\mathrm{Tr}(\hat{\rho}_A \log(\hat{\rho}_A))$ of the reduced density matrix $\hat{\rho}_A = \mathrm{Tr}_B(\hat{\rho})$ is a measure for *bipartite* entanglement [11, 13, 22] between one subsystem $\hat{\rho}_A$ and the rest of the system $\hat{\rho}_B = \mathrm{Tr}_A(\hat{\rho})$. There is

---

[3] The general expression is $\frac{4\Delta \hat{J}_k^2}{N} \geq 1 - \frac{4\langle \hat{J}_k \rangle^2}{N^2}$ [19].

[4] Since criterion (2.15) can be related to a gain in interferometric precision (see Sect. 2.4), it measures the "usefulness" of spin squeezed states as a quantum resource in a known experimental protocol.

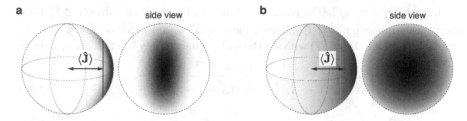

Fig. 2.4 Von Neumann entropy and delocalization of the quantum state. Panel **a** depicts an entangled spin squeezed state on the Bloch sphere. Quantum correlations cause an increased uncertainty in one spin direction which results in a shortening of the effective spin length. This shortening is measured by the linearized von Neumann entropy. Panel **b** shows a non-entangled incoherent mixture. Loss of coherence results also in a shortening of the mean spin length, making it hard to use the von Neumann entropy for our experiments where temperature or environmental noise cause decoherence

no difference on which of the two subsystem S_N is evaluated: $S_N(\hat{\rho}_A) = S_N(\hat{\rho}_B)$. Expanding the von Neumann entropy to first order one obtains the linear entropy:

$$S_N = 1 - \mathsf{Tr}(\hat{\rho}_A^2) \tag{2.22}$$

Taking subsystem A to be a single elementary spin 1/2 particle, S_N can be used to measure entanglement between one Qubit and the rest of the system. The density matrix $\hat{\rho}_A$ can be expressed as a linear combination of Pauli matrices σ_i [23]. If the system is additionally in a symmetric state, the linearized von Neumann entropy can be related to the mean values of the collective spin J [21, 24]:

$$S_N = \frac{1}{2}[1 - \frac{4}{N^2}(\langle \hat{J}_x \rangle^2 + \langle \hat{J}_y \rangle^2 + \langle \hat{J}_z \rangle^2)] \tag{2.23}$$

Figure 2.4 illustrates the linear entropy measure and clarifies its connection to the spread of the state on the Bloch sphere. Since mixed states always have an (incoherently) increased spread it is essential to note its applicability to pure states only. The quantum states realized in our experiments are subject to decoherence making it impossible to apply the linear entropy measure.

Depth of Entanglement

In the context of spin squeezing the *depth of entanglement* has been proposed to quantify entanglement [25] which measures the number of non-separable elementary Qubits. This criterion is valid for incoherent mixtures as well as for pure states making it suitable for our experiments. However, we once again emphasize that there is no clear definition for entanglement among indistinguishable particles. Furthermore, unique entanglement measures for more than two or three particles are still a very active field of research [11].

We review the depth of entanglement criterion here and use the label J for the collective spin of the full system and the label S for subsystems of smaller spin, but not necessarily $S = 1/2$. The basic idea is to find the minimal variance $\Delta \hat{S}_z^2$ for a given mean spin length $\langle \hat{S}_x \rangle$. Combining the inequality $\langle \hat{S}_z^2 \rangle + \langle \hat{S}_y^2 \rangle + \langle \hat{S}_z^2 \rangle \leq S(S+1)$ (which is similar to Eq. 2.17) with the Heisenberg uncertainty limit (2.2) one obtains

$$\Delta \hat{S}_z^2 \geq \frac{1}{2} \left[S(S+1) - \langle \hat{S}_x \rangle^2 - \sqrt{(S(S+1) - \langle \hat{S}_x \rangle^2)^2 - \langle \hat{S}_x \rangle^2} \right] \tag{2.24}$$

as an analytical estimation of the limit.

Numerical calculations allow to set the bound even tighter [25] and a comparison between the numerical results and the analytical formula is shown in Fig. 2.5. From Fig. 2.5 it is obvious that large spins S can be more squeezed than small spins.[5] This implies that a collective spin J composed of k subsystems with spin $S^{(k)}$ can be more squeezed than the individual spins $S^{(k)}$. In other words, one perfectly squeezed large spin J has always lower or equal normalized variance $\Delta \hat{J}_z^2 / J$ for a given normalized mean spin length $\langle \hat{J}_x \rangle / J$ than the sum of the normalized variances of N independent but individually perfectly squeezed smaller spins $S^{(k)}$ for the same normalized mean spin length. Based on these findings the authors of Ref. [25] derive a lower bound for the variance of the collective spin $\Delta \hat{J}_z^2$

$$\Delta \hat{J}_z^2 / NS \geq F_S(\langle \hat{J}_z \rangle / NS) \tag{2.25}$$

where $F_S(.)$ is the minimum for spin S shown in Fig. 2.5.

The interpretation of this result in the case of N spin 1/2 particles is: If one measures the pair $\Delta \hat{J}_z^2 / J$ and $\langle \hat{J}_x \rangle / J$ outside the gray shaded area in Fig. 2.5, entanglement has to be present in the system. Depending on which curve m the measured datapoint falls, the minimal size of the largest non-separable spin has to be $S = m \cdot 1/2$ and the number of these non-separable blocks is N/m.

What happens if N/m is not an integer value? In this case there has to be one or more smaller blocks of entangled (or even non-entangled) particles, causing larger fluctuations than in the case of exactly N/m particles with spin $S = m \cdot 1/2$ since smaller spins cause larger fluctuations. In order to explain the observed data point, the largest entangled block has to be even greater than m. To summarize, minimally m entangled Qubits are detected if the measured datapoint falls on the curve for $S = m \cdot 1/2$.

2.4 Entangled Interferometry

Entanglement in collective spin systems is not only interesting from a conceptual perspective but it has also been shown to provide a useful quantum resource. In 1994 Wineland et al. [2] pointed out, that in particular spin squeezed states can be used to overcome the standard quantum limit in metrology.

---

[5] As already mentioned a spin $S = 1/2$ can not be squeezed at all.

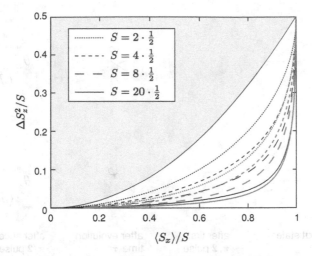

Fig. 2.5 Depth of entanglement. The figure shows the minimal allowed normalized variance $\Delta \hat{S}_z^2$ for a given mean spin length $\langle \hat{S}_x \rangle$ depending on the total spin S (different line styles). The black lines are the numerical result taken from reference [25] (Fig. 2.1) while the gray lines show the analytic approximation (2.24) which we use later in this thesis. The spin length S is written as $S = m \cdot 1/2$ in order to emphasize the minimal non-separable block size m of the density matrix in the case of Qubits as elementary spins. The gray area correspond to pairs of $\langle \hat{S}_x \rangle / S$ and $\Delta \hat{S}_z^2 / S$ for which no entanglement is detected in the system

2.4.1 Precision Limits in Ramsey Interferometry

The term *Ramsey interferometry* [26, 27] is used most often for atomic interferometers based on internal states. Prominent applications are the definition of the time standard [28] or high precision magnetometry [29]. However the scheme is more general and applies also to atom interferometers where the two states are implemented using external degrees of freedom. These interferometers allow for example for high precision inertia measurements of gravity or rotation [30–32]. The optical counterpart of Ramsey interferometry is a Mach–Zehnder interferometer and the analogy is further discussed in Sect. 4.7 .

The Ramsey Interferometric Sequence

In order to develop an intuitive understanding for the precision limit in interferometry we discuss the implementation of a typical Ramsey interferometer and visualize the protocol schematically on Bloch spheres (Fig. 2.6a). A Ramsey atom interferometer conceptually consists of at least three building blocks, two beamsplitters and an evolution time in between. The first beamsplitter, which corresponds to a unitary rotation on the Bloch sphere around an axis in the equatorial plane, is used to generate a coherent superposition of the two quantum states. Assuming only one input port to be populated the output is usually a collective spin state with the mean spin pointing onto the equator. A fixed time τ of free evolution follows during which a relative

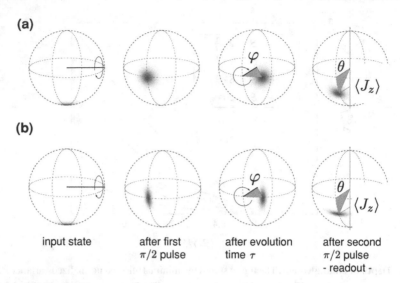

(a)

(b)

input state   after first   after evolution   after second
                     $\pi/2$ pulse   time τ $\pi/2$ pulse
                                                                   - readout -

Fig. 2.6 Schematic representation of Ramsey interferometry on the Bloch sphere. Part **a** shows the standard Ramsey protocol represented on the Bloch sphere. Beamsplitters correspond to rotations of the quantum state around an axis in the equatorial plane as indicated by the circular shaped arrows. The sequence is described in detail in the main text. Panel **b** shows a similar protocol but after the first "magic" beamsplitter a spin squeezed state emerges which propagates through the interferometer resulting in degreased occupation number uncertainty at the readout. Section 4.7 of this thesis describes the concrete implementation of this "magic"—non-linear—beamsplitter

phase φ between the two modes accumulates (corresponding to a longitudinal rotation on the Bloch sphere). Depending on the kind of interferometer this phase is due to differential energy shifts between the states or due to effective path length differences to be measured [30]. Since the angle in longitudinal direction on the Bloch sphere φ is usually not directly observable, a second beamsplitter is necessary. This beamsplitter implements another unitary rotation around an axis in the equatorial plane shifted by 90° with respect to the first beamsplitter in order to translate the longitudinal angle to a polar angle θ. The readout of the interferometer is done by detection of the population difference J_z of the two output ports, from which the relative phase φ can be deduced. The resulting sinusoidal variation of the population difference $\langle \hat{J}_z \rangle$ versus acquired relative phase φ is commonly called a *Ramsey fringe*.

Quantifying Interferometric Precision

Taking finite environmental noise into account, the sensitivity of the interferometer to small phase shifts

$$\Delta\varphi^{-1} = \left(\frac{\Delta \hat{J}_z}{\frac{\partial \langle \hat{J}_z \rangle}{\partial \varphi}} \right)^{-1} \tag{2.26}$$

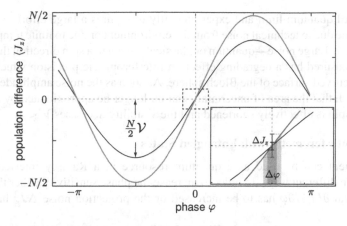

Fig. 2.7 Precision limit in Ramsey interferometry. We compare schematically the phase estimation precision in Ramsey interferometry using a coherent spin state (*gray*) and a spin squeezed state (black). The main figure shows a Ramsey fringe whose visibility $\mathcal{V}$ is maximal for a coherent spin state ($\mathcal{V} = 1$) but smaller for a spin squeezed state. Nevertheless the phase precision for a squeezed state outperforms the precision obtained for classical interferometer as shown in the zoom around the zero crossing. The projection noise is suppressed for the spin squeezed state such that the ratio of projection noise and slope of the Ramsey fringe is smaller by a factor ξ_S compared to the standard quantum limit, which explains the gain in interferometric precision

depends on the mean phase $\langle \varphi \rangle$ and is determined by the projection noise $\Delta \hat{J}_z$ and the slope of the Ramsey fringe $\partial \langle \hat{J}_z \rangle / \partial \varphi$. The point of maximum sensitivity is reached where the mean population difference is zero and the slope is maximal $(\partial \langle \hat{J}_z \rangle / \partial \varphi)_{\text{max}} = \mathcal{V} N / 2$. The visibility $\mathcal{V}$ measures the mean spin length $\langle \mathbf{J} \rangle = \mathcal{V} N / 2$. Figure 2.7 illustrates the phase sensitivity of a Ramsey interferometer.

The amount of precision gain (or loss) relative to standard quantum limit is given by ξ_R^{-2} or equivalently by $\left(\xi_S^2 \right)^{-1}$. The measure can be expressed in visibility $\mathcal{V}$ and spin noise in J_z direction at readout $\Delta \hat{J}_z^2$:

$$\xi_S^2 = \frac{4 \Delta \hat{J}_z^2}{\mathcal{V}^2 N} \tag{2.27}$$

The absolute phase uncertainty–measured as the root mean square deviation is:

$$\Delta \varphi = \xi_S \frac{1}{\sqrt{N}} \tag{2.28}$$

Spin squeezed states feature reduced noise in one of the spin directions but excess noise in another direction can be present either due to a non-Heisenberg limited quantum state or due to an incoherent mixture of several quantum states. The former might limit precision in standard Ramsey interferometry, but specific correlated quantum states enable even enhanced interferometric precision in a generalized interferometer [33]. The latter is easily limiting interferometric precision at a level above the standard quantum limit and experimentally it requires a large effort to prevent decoherence due to technical noise from the environment or due to finite temperature in the system. Large noise—quantum or classical—even in a spin direction that is not directly measured has a degrading effect on interferometric precision which arises due to the curved surface of the Bloch sphere. As soon as the noise amplitude is large enough such that the area of uncertainty can no longer be approximated by a plane, the mean spin is effectively shortened and the visibility decreases $\mathcal{V} < 1$.

Ramsey Interferometry with Entangled States

Entanglement can be used as a quantum resource in a Ramsey interferometric sequence in different ways. In order to increase the phase sensitivity either the slope of the signal $\partial\langle \hat{J}_z \rangle / \partial\varphi$ has to be increased or the projection noise $\Delta \hat{J}_z^2$ has to be decreased.

Slope increase can be reached by *Schrödinger cat* type entanglement which involves maximally entangled states that are very fragile to decoherence. Therefore they have been realized so far with very few particles only [34–36].

Spin squeezing aims to decrease the projection noise. This is possible in gradual steps meaning that depending on the amount of spin squeezing the precision is gradually increased. Therefore—at least for moderate levels of spin squeezing—these states are less fragile and they have been realized with a large number of particles but only with a relatively small squeezing factor [37–46]. Ramsey interferometry with spin squeezed states is schematically depicted in Fig. 2.6b where a "magic" beamsplitter produces an entangled state. Interferometric sensitivity for a coherent spin state and a spin squeezed state is compared in Fig. 2.7. For the spin squeezed state the decreased quantum fluctuations $\Delta \hat{J}_z$ reduce the projection noise while the increased fluctuations $\Delta \hat{J}_y$ cause a slight decrease of the mean spin length and therefore of the visibility of the Ramsey fringe. Nevertheless, the ratio of projection noise and slope of the Ramsey fringe—and therefore the phase sensitivity—is increased.

2.4.2 Heisenberg Limit in Quantum Metrology

The ultimate limit for metrologic precision is the *Heisenberg limit* [47], where the phase estimation error $\Delta\varphi$ is given by

$$\Delta\varphi = \frac{1}{N} \tag{2.29}$$

for N resources used in a single measurement. This fundamental limit can—up to a constant numerical factor in the order of unity—in principle be reached with both approaches mentioned above—Schrödinger cat type entanglement or spin squeezing.

Schrödinger Cats and Metrology

In the context of quantum metrology the Schrödinger cat state is frequently called a NOON state [48]. Its name originates from its form in Fock states basis:

$$|NOON\rangle = (|N, O\rangle + e^{i\varphi_N}|O, N\rangle)/\sqrt{2} \qquad (2.30)$$

It is a coherent superposition of all atoms in state $\hat{a}$ and zero atoms in state $\hat{b}$ and vice versa. In spin representation the NOON state is the superposition of the two maximal Dicke states:

$$|NOON\rangle = (|J, -J\rangle + e^{i\varphi_N}|J, J\rangle)/\sqrt{2} \qquad (2.31)$$

The increase of the signal slope for a NOON state is obvious since the phase acquired between the two components $\varphi_N = N\varphi$ is N times larger than for a coherent spin state [9, 49, 50]. Experimentally it is important to note that the readout of the interferometer can not be realized by measuring $\langle \hat{J}_z \rangle$. The reason is the vanishing mean spin length $\langle \hat{J}_x \rangle$ of this state. It has been shown that the parity of the state is a useful experimental observable to make use of NOON states in interferometry and to reach the Heisenberg limit [34, 50].

Spin Squeezed States

Spin squeezed states allow to ask for the best achievable interferometry gain demanding a finite mean spin length such that standard interferometric readout can be used.

The optimum ξ_R for a given mean spin length was found numerically in Ref. [25] and for rather small spins it is shown in Fig. 2.5. An experimental protocol to generate spin squeezed states close to the Heisenberg limit was proposed in Ref. [52].

Other Types of Quantum Correlated States

Recently it has been pointed out that the *Fisher information* is the most general criterion to measure phase sensitivity since it saturates the Quantum Cramer-Rao bound [33, 53]. Calculating the Fisher information for a coherent spin state state evolving under the non-linear Hamiltonian $\hat{H} = \chi \hat{J}_z^2$, where χ parametrizes the nonlinearity, Pezzé and Smerzi recovered Heisenberg limit like scaling for the phase precision [33]. The quantum state here is neither necessarily a NOON state nor a coherently spin squeezed state. However standard interferometric readout can not be used to extract the phase information and a new type of Bayesian readout has to be employed which was experimentally demonstrated in [53].

References

1. Kitagawa M, Ueda M (1993) Squeezed spin states. Phys Rev A 47:5138–5143
2. Wineland D, Bollinger J, Itano W, Heinzen D (1994) Squeezed atomic states and projection noise in spectroscopy. Phys Rev A 50:67–88

3. Sørensen AS, Duan L, Cirac J, Zoller P (2001) Many-particle entanglement with Bose–Einstein condensates. Nature 409:63–6
4. Metcalf H, Van der Straten P (1999) Laser cooling and trapping. Springer, New York
5. Sakurai J (1994) Modern quantum mechanics. Addison-Wesley, Reading
6. Arecchi FT, Courtens E, Gilmore R, Thomas H (1972) Atomic coherent states in quantum optics. Phys Rev A 6:2211–2237
7. Radcliffe JM (1971) Some properties of coherent spin states. J Phys A Gen Phys 4:313–323
8. Zhang W-M, Feng DH, Gilmore R (1990) Coherent states: theory and some applications. Rev Mod Phys 62:867–927
9. Giovannetti V, Lloyd S, Maccone L (2004) Quantum-enhanced measurements: beating the standard quantum limit. Science 306:1330–1336
10. Lee CT (1984) Q representation of the atomic coherent states and the origin of fluctuations in superfluorescence. Phys Rev A 30:3308–3310
11. Amico L, Fazio R, Osterloh A, Vedral V (2008) Entanglement in many-body systems. Rev Mod Phys 80:517
12. Horodecki R, Horodecki P, Horodecki M, Horodecki K (2009) Quantum entanglement. Rev Mod Phys 81:865
13. Plenio MB, Virmani S (2007) An introduction to entanglement measures. Quantum Inf Comput 7:1
14. Benatti F, Floreanini R, Marzolino U (2010) Sub-shot-noise quantum metrology with entangled identical particles. Ann Phys 325:924
15. Tóth G, Knapp C, Gühne O, Briegel HJ (2007) Optimal spin squeezing inequalities detect bound entanglement in spin models. Phys Rev Lett 99:250405
16. Tóth G, Knapp C, Gühne O, Briegel HJ (2009) Spin squeezing and entanglement. Phys Rev A 79:042334
17. Korbicz J et al (2006) Generalized spin-squeezing inequalities in N-qubit systems: theory and experiment. Phys Rev A 74:52319
18. Korbicz JK, Cirac JI, Lewenstein M (2005) Erratum spin squeezing inequalities and entanglement of N qubit states. Phys Rev Lett 95:259901
19. Korbicz JK, Cirac JI, Lewenstein M (2005) Spin squeezing inequalities and entanglement of N qubit states. Phys Rev Lett 95:120502
20. Wang X, Sanders B (2003) Spin squeezing and pairwise entanglement for symmetric multiqubit states. Phys Rev A 68:12101
21. Chaudhury S, Smith A, Anderson BE, Ghose S, Jessen PS (2009) Quantum signatures of chaos in a kicked top. Nature 461:768–771
22. Bennett CH, DiVincenzo DP, Smolin JA, Wootters WK (1996) Mixed-state entanglement and quantum error correction. Phys Rev A 54:824–3851
23. Cohen-Tannoudji C, Diu B, Laloe F (2005) Quantum mechanics. Wiley-VCH, New York
24. Ghose S, Stock R, Jessen P, Lal R, Silberfarb A (2001) Chaos, entanglement, and decoherence in the quantum kicked top. Phys Rev A 78:042318
25. Sørensen AS, Mølmer K (1949) Entanglement and extreme spin squeezing. Phys Rev Lett 86:4431–4434
26. Ramsey NF (1949) A new molecular beam resonance method. Phys Rev 76:996
27. Ramsey NF (1950) A molecular beam resonance method with separated oscillating fields. Phys Rev 78:695–699
28. Santarelli G et al (1999) Quantum projection noise in an atomic fountain: a high stability cesium frequency standard. Phys Rev Lett 82:4619–4622
29. Wasilewski W, Jensen K, Krauter H, Renema JJ, Polzik ES (2010) Quantum noise limited and entanglement assisted magnetometry. Phys Rev Lett 104:133601
30. Cronin AD, Schmiedmayer J, Pritchard DE (2009) Optics and interferometry with atoms and molecules. Rev Mod Phys 81:1051
31. Kasevich M, Chu S (1992) Measurement of the gravitational acceleration of an atom with a light-pulse atom interferometer. Appl Phys B 54:321–332

32. Gustavson TL, Bouyer P, Kasevich MA (1997) Precision rotation measurements with an atom interferometer gyroscope. Phys Rev Lett 78:2046–2049
33. Pezzé L, Smerzi A (2009) Entanglement, nonlinear dynamics, and the Heisenberg limit. Phys Rev Lett 102:100401
34. Leibfried D et al (2004) Toward Heisenberg-limited spectroscopy with multiparticle entangled states. Science 304:1476–1478
35. Roos CF, Chwalla M, Kim K, Riebe M, Blatt R (2006) 'Designer atoms' for quantum metrology. Nature 443:316
36. Nagata T, Okamoto R, O'Brien JL, Sasaki K, Takeuchi S (2007) Beating the standard quantum limit with four-entangled photons. Science 316:726–729
37. Meyer V et al (2001) Experimental demonstration of entanglement-enhanced rotation angle estimation using trapped ions. Phys Rev Lett 86:5870–5873
38. Schleier-Smith MH, Leroux ID, Vuletic V (2010) Reduced-quantum-uncertainty states of an ensemble of two-level atoms. Phys Rev Lett 104:73604
39. Leroux ID, Schleier-Smith MH, Vuletic V (2010) Implementation of cavity squeezing of a collective atomic spin. Phys Rev Lett 104:73602
40. Fernholz T et al (2008) Spin squeezing of atomic ensembles via nuclear-electronic spin entanglement. Phys Rev Lett 101:073601
41. Kuzmich A, Mandel L, Bigelow NP (2000) Generation of spin squeezing via continuous quantum nondemolition measurement. Phys Rev Lett 85:1594–1597
42. Hald J, Sørensen JL, Schori C, Polzik ES (1999) Spin squeezed atoms: a macroscopic entangled ensemble created by light. Phys Rev Lett 83:1319–1322
43. Appel J et al (2009) Mesoscopic atomic entanglement for precision measurements beyond the standard quantum limit. Proc Natl Acad Sci USA 106:10960–10965
44. Goda K et al (2008) A quantum-enhanced prototype gravitational-wave detector. Nat Phys 4:472–476
45. Vahlbruch H et al (2008) Observation of squeezed light with 10 dB quantum-noise reduction. Phys Rev Lett 100:033602
46. Estève J, Gross C, Weller A, Giovanazzi S, Oberthaler M K (2008) Squeezing and entanglement in a Bose–Einstein condensate. Nature 455:1216–1219
47. Giovannetti V, Lloyd S, Maccone L (2006) Quantum metrology. Phys Rev Lett 96:010401
48. Lee H, Kok P, Dowling J (2003) A quantum Rosetta stone for interferometry. J Mod Opt 49:2325–2338
49. Bouyer P, Kasevich MA (1997) Heisenberg-limited spectroscopy with degenerate Bose–Einstein gases. Phys Rev A 56:R1083–R1086
50. Dowling JP (1998) Correlated input-port, matter-wave interferometer: Quantum-noise limits to the atom-laser gyroscope. Phys Rev A 57:4736–4746
51. Bollinger JJ, Itano WM, Wineland DJ, Heinzen DJ (1996) Optimal frequency measurements with maximally correlated states. Phys Rev A 54:R4649–R4652
52. Pezzé L, Collins LA, Smerzi A, Berman GP, Bishop AR (2005) Sub-shotnoise phase sensitivity with a Bose–Einstein condensate Mach–Zehnder interferometer. Phys Rev A 72:043612
53. Pezzé L, Smerzi A, Khoury G, Hodelin JF, Bouwmeester D (2007) Phase detection at the quantum limit with multiphoton Mach–Zehnder interferometry. Phys Rev Lett 99:223602

Chapter 3
Squeezing Two Mean Field Modes of a Bose–Einstein Condensate

Bose–Einstein condensation has been predicted in 1924/1925 by Satyendra Nath Bose and Albert Einstein [1–3]. The Nobel prize 2001 was awarded to Eric A. Cornell, Wolfgang Ketterle and Carl E. Wieman for the first experimental observation of Bose–Einstein condensation in dilute gases of laser cooled alkali atoms in 1995 [4–8]. Almost 15 years later a whole new sub field of atomic physics developed dealing with Bose–Einstein condensates and degenerate Fermi gases. A lot of effort has been made, both experimentally and theoretically, to explore the basic physics of ultracold quantum degenerate gases [9–11]. Extraordinary experimental control over the trapped quantum gases and the possibility to measure and adjust almost all relevant parameters directly (e.g. interaction strength, relative phases, ...) opens up a new route in atomic physics. The quantum gases can be used to engineer specific Hamiltonians that map for example to problems in solid state physics where some measurements are hard to perform and many parameters are not controllable. Ultracold quantum gases are promising candidates for quantum simulators of solid state systems [12–14]. In the field of quantum metrology degenerate gases have been proposed to be one experimental system that allows for a precision beyond the "classical" projection noise limit in atom interferometry. Controllable many-body entanglement can be used as a resource to beat the standard quantum limit [15–21].

In this chapter we focus on Bose–Einstein condensates in double- and few-well potentials and in particular on the experimental observation of spin squeezing type many-body entanglement among them. The mechanism of squeezing generation is explained within a two-mode approximation and limits on the observed spin squeezing due to finite temperature and environmental noise are discussed.

C. Groß, *Spin Squeezing and Non-linear Atom Interferometry with Bose–Einstein Condensates*, Springer Theses, DOI: 10.1007/978-3-642-25637-0_3, © Springer-Verlag Berlin Heidelberg 2012

3.1 Bose–Einstein Condensates in Double-Well Potentials: Mean Field and Beyond

3.1.1 Basic Concepts of Bose–Einstein Condensation

In this section we discuss some basic principles of Bose–Einstein condensation in dilute alkali gases necessary to understand the main part of this thesis—coherent spin squeezing in Bose–Einstein condensates and its limits. We follow the arguments given in the books [9, 10].

Bose–Einstein condensation is a quantum statistical effect that occurs for non- or weakly interacting Bosons. The occupation number of a single particle state n_i obeys the Bose–Einstein statistics $n_i = [e^{(\epsilon_i - \mu)/k_B T} - 1]^{-1}$, where k_B is Bolzmann's constant and ϵ_i the single particle eigenenergy. For high temperatures $T \gg T_C$ the chemical potential μ is much lower than the single particle ground state eigenenergy ϵ_0. With decreasing temperature the phase space density $\tilde{\rho} = n\lambda_T^3$ and simultaneously the chemical potential μ rises. Here n is the atomic density and $\lambda_T^3 = (2\pi\hbar^2/mk_B T)^{3/2}$ is the cubic thermal de Broglie wavelength.[1] When the phase space density $\tilde{\rho}$ exceeds a critical value in the order of unity,[2] μ approaches ϵ_0 and the ground state becomes macroscopically occupied $n_0 \approx N$. This is the mechanism of Bose–Einstein condensation. The functional dependence of the fraction of atoms in the condensate n_0/N as a function of temperature is determined by the density of states which is given by the dimensionality of the system and the spatial trapping potential V_{ext}. For a three dimensional harmonic trap n_0 follows from:

$$\frac{n_0}{N} = 1 - \left(\frac{T}{T_C}\right)^3 \tag{3.1}$$

In alkali vapors with typical densities between 10^{13} and $10^{15}\,\mathrm{cm}^{-3}$ the critical temperature T_C is in the 100 nK to few μK regime.

Interacting Bosons

Alkali vapors are not exactly ideal gases but they are weakly interacting dilute gases. Dilute means that the gas parameter is much smaller than unity $na^3 \ll 1$, where a is the s-wave scattering length describing the interactions among the atoms as contact interactions at low temperatures. The ideals gas formalism remains approximately valid, but the interparticle interaction causes a modification of the single particle eigenenergies. The condensate emerges in the lowest collective, *mean field* state. The theoretical description of weakly interacting dilute Bose gases was introduced 1947 by Bogoliubov. The key idea is to replace the annihilation (and creation) operators $\hat{a}_0$ for the macroscopically occupied ground state by a complex number $\hat{a}_0 \rightarrow \sqrt{n_0}e^{i\varphi_0}$.

---

[1] $2\pi\hbar$ is Planck's constant and m *the atomic mass.*
[2] The exact value depends on the density of states.

Starting with the exact Hamiltonian of the system expressed in field operators $\hat{\Psi} = \sum_i \phi_i \hat{a}_i$

$$\hat{H} = \int d\mathbf{r} \hat{\Psi}^\dagger \left(-\frac{\hbar \nabla}{2m} + V_{ext} \right) \hat{\Psi} + \frac{1}{2} \int d\mathbf{r} d\mathbf{r}' \hat{\Psi}^\dagger \hat{\Psi}^{\dagger'} V(\mathbf{r} - \mathbf{r}') \hat{\Psi} \hat{\Psi}' \qquad (3.2)$$

one way to obtain the *Gross–Pitaevskii equation* is to minimize the grand canonical potential $\hat{\Omega} = \hat{H} - \mu \hat{N}$ under the Bogoliubov approximation neglecting all states but the ground state.[3] Minimization is done with respect to the condensate wavefunction Ψ_0 meaning $\frac{\partial \Omega[\Psi_0, \Psi_0^*]}{\partial \Psi_0^*} = 0$ and the Gross–Pitaevskii equation found 1961 independently by Pitaevskii and Gross reads:

$$i\hbar \frac{\partial}{\partial t} \Psi_0(\mathbf{r}, t) = \left(-\frac{\hbar \nabla}{2m} + V_{ext}(\mathbf{r}, t) + g |\Psi_0(\mathbf{r}, t)|^2 \right) \Psi_0(\mathbf{r}, t) \qquad (3.3)$$

Here $g = \frac{4\pi \hbar^2 a}{m}$ is the coupling constant proportional to the s-wave scattering length a. Equation (3.3) describes a weakly interaction Bose–Einstein condensate in the mean field limit. It reveals the ground state of the system Ψ_0, but it does not tell anything about fluctuations of the system.

3.1.2 A Bosonic Josephson Junction with Ultracold Atoms

Some Aspects of Josephson Junctions in Ultracold Gases and in Solid State Systems

A Bose–Einstein condensate in an external double-well potential models a Josephson junction in solid state systems [22]. However one big difference is that the trapped ultracold gas is a closed system meaning the wavefunction vanishes for large distances or equivalently the number of Bosons in the system is fixed. For a solid state junction the system is coupled to the environment by current carrying wires resulting in a non-fixed number of cooper pairs in the system. Due to these differences the experimental observables to characterize the state of the system differ. In solid state systems transport properties like the current through the junction or the voltage across it can be measured, however there is no direct way to "look at" the spatial probability distribution of the cooper pairs. In the case of ultracold gases however the distribution of atoms, their number and the relative phase between the two wells can be directly measured.

Josephson Junctions for Ultracold Bosons—Experimental setup

An accurate description of our Bose–Einstein condensation apparatus can be found in former Ph.D. thesis from our group [23–25] such that only the experimental parts essential for the experiments presented in this thesis are discussed here. A red

---

[3] ϕ_i is the normalized ith eigenfunction of the single particle Hamiltonian, $V(\mathbf{r} - \mathbf{r}')$ is the interatomic interaction potential, later approximated as a contact interaction $V(\mathbf{r} - \mathbf{r}') \propto \delta(\mathbf{r} - \mathbf{r}')$.

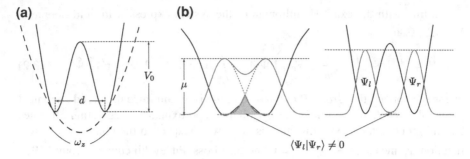

Fig. 3.1 Trap geometry of the double-well potential. **a** illustrates the most important trap parameters which are the well distance $d = 5.7\,\mu m$, the barrier height V_0 and the longitudinal trap frequency of the dipole trap ω_z. **b** The chemical potential μ and the barrier height V_0 define the wavefunction overlap $\langle \Psi_l | \Psi_r \rangle$ between the *left* and the *right* mode. The overlap defines the Josephson coupling E_J between the modes which can be controlled by the barrier height V_0

detuned optical dipole trap [26] with a wavelength of 1,064 nm is used to generate the external trapping potentials for the Bose–Einstein condensate of [87] Rubidium atoms in the $F = 2$, $m_F = 2$ hyperfine state [27]. Harmonic approximation of the trapping potential around the potential minimum reveals the trap frequencies ω_i of the bare dipole trap. The transversal frequencies in our tightly focussed single beam trap[4] are $\omega_x = \omega_y = 2\pi \cdot 425$ Hz and the longitudinal frequency is $\omega_z = 2\pi \cdot 20$ Hz. A second orthogonal trapping beam can be used additionally to increase the longitudinal trapping frequency continuously up to $\omega_z = 2\pi \cdot 70$ Hz while the two transversal frequencies remain almost unchanged. An one dimensional optical lattice—generated using a laser with a wavelength of 843 nm—superimposed in longitudinal direction allows for splitting of the trap into two or more wells depending on the longitudinal confinement ω_z. The periodicity of the optical lattice is set to $d = 5.7\,\mu m$ by choosing an angle of 8.5° between the two lattice beams. The height of the potential barrier $V_0 \propto I_{latt}$ between the wells can be accurately controlled by adjusting the intensity I_{latt} of the optical lattice beams.[5] Further technical details of the lattice setup and its calibration can be found in reference [23]. Figure 3.1a schematically shows the double-well trap geometry and in Fig. 3.10 later in this thesis the laser beam configuration can be found.

The Josephson Hamiltonian

The quantum state of a Josephson junction is governed by two competing processes. Overlap of the wavefunctions in the barrier region (Fig. 3.1b) results in a tunnel contact with a Josephson energy E_J and therefore in a finite probability for the atoms (or cooper pairs in solid state junctions) to cross the barrier. In the bosonic Josephson junction realized in our lab this Josephson energy is easily adjustable throughout the experiments since the wavefunction overlap depends on the height

---

[4] The beam waist is 5.1 μm.

[5] The barrier height is tunable between $V_0 = 2\pi \cdot 250$ Hz and $V_0 = 2\pi \cdot 3,000$ Hz.

of the barrier V_0. On the other hand there is a finite effective interaction energy E_C between the bosons which is in the solid state case governed by the junction capacitance and in the atomic case it is given by the elastic interactions between the atoms in the individual wells. The Josephson Hamiltonian

$$H = \frac{1}{2} E_C n^2 - E_J \cos(\varphi) \tag{3.4}$$

describes this situation with occupation number difference n and relative phase φ between the two wavefunctions.[6]

From here on we focus on an atomic Josephson contact, where the two wavefunctions are the mean field orbitals in the left $\Psi_l = \sqrt{n_l(\mathbf{r})}e^{i\varphi_l}$ and the right $\Psi_r = \sqrt{n_r(\mathbf{r})}e^{i\varphi_r}$ well of the double-well potential with occupation number difference $n = (n_l - n_r)/2$ and relative phase $\varphi = \varphi_l - \varphi_r$.[7] The localized wavefunctions can be expressed as a linear combination of the lowest energy symmetric Ψ_S and antisymmetric Ψ_A solution of the Gross–Pitaevskii equation (3.3):

$$\Psi_{l,r} = \frac{1}{\sqrt{2}}(\Psi_S \pm \Psi_A) \tag{3.5}$$

Charging Energy E_C and Josephson Energy E_J

For a fixed total number of atoms $N = n_l + n_r$ the energies E_C and E_J define the properties of the system. The *Charging energy* E_C is given by the derivative of the chemical potential with respect to the atom number [28]

$$E_C = \frac{\partial \mu}{\partial n_l} = \frac{\partial \mu}{\partial n_r} \tag{3.6}$$

which holds for approximately equal population of the two modes. It is often necessary to estimate the order of magnitude of E_C which, as a rule of thumb, is μ/N. The *Josephson energy* E_J can be calculated using two different approaches. In the regime where the barrier height V_0 between the two wells is greater than the chemical potential μ, E_J is given by the energy difference between the symmetric and antisymmetric mean field orbitals $E_J \propto \mu_A - \mu_S$ [29, 30]. This is very intuitive due to the similarity to a standard two-level atom coupled to an electromagnetic radiation field, where the splitting of the dressed states is the Rabi coupling between the two levels [31]. But one has to be careful with this analogy since interactions alter the properties of the Josephson junction (see Sect. 3.1.3).

The second approach to calculate E_J is valid in a larger range of parameters especially for $V_0 \lesssim \mu$. In reference [28] an analogy of the bosonic Josephson junction to a capacitor in classical electrodynamics is drawn. The fictitious dielectric in the

---

[6] A dependence of $E_J(n)$ from the occupation number difference is omitted here, i.e. we assume $n \ll N$. This term is included in the discussion presented in Sect. 3.1.3.

[7] $n_{l,r}(\mathbf{r})$ is the atomic density of the left (right) mode and $n_{l,r} = \int_{-\infty,0}^{0,\infty} d\mathbf{r} n_{l,r}(\mathbf{r})$ the mode occupation number.

capacitor is nonuniform and its distribution is given by atomic density $n(\mathbf{r})$. Based on this analogy $E_J = \frac{\hbar^2}{m} C$ is calculated from the capacitance of this capacitor C which is bounded from below and above

$$\int \frac{\mathrm{d}x\mathrm{d}y}{\int_{z_l}^{z_r} \mathrm{d}z \, n(x,y,z)^{-1}} \le C \le \left[\int_{z_l}^{z_r} \frac{\mathrm{d}z}{\int \mathrm{d}x\mathrm{d}y \, n(x,y,z)} \right]^{-1} \qquad (3.7)$$

where the two double-well potential minima are located at z_l and z_r.

Numerical Calculation of E_C and E_J

Both parameters E_C and E_J can be calculated from the mean field wavefunction obtained from the Gross–Pitaevskii equation (3.3). In our group a numerical code exists to solve the equation in three dimensions by a split step Fourier transform algorithm [25].

In Fig. 3.2 the two approaches to calculate the Josephson energy E_J are compared. In a harmonic trap the fastest possible tunneling time is bounded by the inverse trap frequency ω_z^{-1} since the trap frequency sets the minimal time for the atoms to move from one side of the trap to the other. Figure 3.2b shows that the "capacitor" method to calculate E_J gives correct results in the regime $V_0 < \mu$ since the tunnel frequency ω_{pl} approaches the trap frequency ω_z in the limit of vanishing barrier height (see Sect. 3.2.1 for the definition of ω_{pl}).

Figure 3.3 shows E_C and E_J calculated for different barrier heights V_0 separating the two wells. The calculation is done for the dipole trap parameters valid for our double-well experiments and 1,600 atoms in total. With increasing barrier height, E_J drops a few orders of magnitude while E_C stays constant within a factor of two. This identifies E_J as the main control parameter for our system.

3.1.3 Rabi, Josephson and Fock: Different Regimes of a Josephson Junction

The Josephson Junction in Two-Mode Approximation

The mean field treatment of the Josephson junction discussed above is not sufficient to explain fluctuations in the occupation number difference n and the relative phase φ. We employ a two-mode description with constant parameters obtained from the mean field model to describe these fluctuations.

Within the two-mode approximation $\hat{\Psi} = \phi_l \hat{a}_l + \phi_r \hat{a}_r$ the Josephson Hamiltonian (3.4) can be derived from the general Hamiltonian (3.2). As a first result the two site *Bose-Hubbard Hamiltonian* is obtained [32]

$$H_{BH} \approx \frac{K}{8}(\hat{a}_l^\dagger \hat{a}_l - \hat{a}_r^\dagger \hat{a}_r)^2 - \frac{\Delta E}{2}(\hat{a}_l^\dagger \hat{a}_l - \hat{a}_r^\dagger \hat{a}_r) - \frac{J}{2}(\hat{a}_l^\dagger \hat{a}_r + \hat{a}_r^\dagger \hat{a}_l) \qquad (3.8)$$

where the large occupation number per site leads to renormalized (compared to single particle parameters) onsite interactions $K = E_C$ and hopping $J = 2E_J/N$ [33].

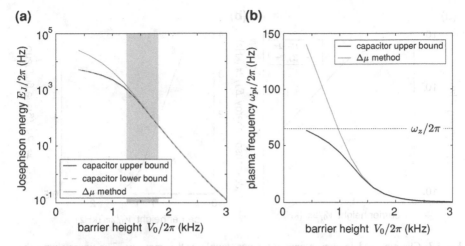

Fig. 3.2 Two methods to calculate the Josephson energy. **a** The Josephson energy E_J, calculated using the two methods detailed in the main text is plotted versus the height of the potential barrier. The two bounds for the capacitance are not distinguishable on this scale, while for $V_0 < \mu$ the deduced Josephson energy differs significantly from the value obtained with the "$\Delta\mu$" method. The *gray shaded area* gives the range of the chemical potential μ which increases with increasing barrier since the number of atoms $N = 2,000$ is chosen constant while the onsite confinement rises. The calculations were done for a double-well trap with underlying frequencies of the dipole trap of $\omega_x = \omega_y = 2\pi \cdot 420$ Hz and $\omega_z = 2\pi \cdot 65$ Hz. **b** shows the resulting plasma frequency ω_{pl}—the tunneling rate (see Sect. 3.2.1 for more details) between the two wells. The *dashed line* indicates the longitudinal dipole trap frequency, the maximum possible tunnel frequency in our setup. It is obvious that the "$\Delta\mu$" method violates this bound identifying the capacitor method as the correct way to calculate E_J for low potential barriers

These parameters are calculated from the mean field wavefunctions as described in the previous section. The term proportional to ΔE describes a possible differential energy shift between the two modes. In Eq. (3.8) higher order terms like pair tunneling are neglected [30, 34].

The Bose-Hubbard Hamiltonian can be expressed in phase $\hat{\varphi}$ and number difference $\hat{n}$ operators which reveals the Josephson Hamiltonian. Definition of the phase operator is not straightforward, however it can be done in the limit of large atom number N [32]. Phase and number operator fulfill the canonical commutation relation $[\hat{\varphi}, \hat{n}] = i$. The Josephson Hamiltonian is then given by:

$$\hat{H}_J = -\Delta E \hat{n} + \frac{E_C}{2}\hat{n}^2 - E_J \sqrt{1 - \frac{4\hat{n}^2}{N^2}} \cos\hat{\varphi} \tag{3.9}$$

Another way describing beyond mean field effects which provides an intuitive picture in many cases is to rewrite the two-mode Bose-Hubbard Hamiltonian using the *Schwinger spin representation* introduced in Sect. 2.1 [32]:

$$H = -\Delta E \hat{J}_z + \frac{E_C}{2}\hat{J}_z^2 - \frac{2E_J}{N}\hat{J}_x \tag{3.10}$$

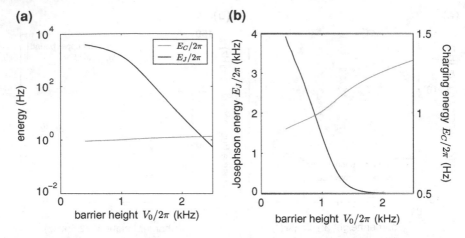

Fig. 3.3 Charging and Josephson energy for our double-well parameters. The figure shows the dependence of the Charging energy E_C and the Josephson energy E_J from the barrier height V_0. On the *left* logarithmic plotting was chosen which reveals an exponential change in E_J for $V_0 > \mu$. The linear double axis plot on the *right hand side* is more useful to highlight the range $400\,\text{Hz} \lesssim V_0 \lesssim 1,700\,\text{Hz}$ where our experiments are carried out

Connected to this spin model the parameters $\chi = E_C/2$ accounting for the nonlinearity and the Rabi frequency $\Omega = 2E_J/N$ describing the coupling of the two modes on the single particle level are commonly found in the literature.

In the limit where Eq. (3.9) is valid it is equivalent to Eq. (3.10) meaning there is a connection between the spin components on one side and the occupation number difference $\hat{n}$ and the relative phase $\hat{\varphi}$ on the other side [32]. For $n \ll N$ the translation between these variables is

$$\hat{J}_x \approx N \cos(\hat{\varphi})/2$$
$$\hat{J}_y \approx N \sin(\hat{\varphi})/2 \qquad (3.11)$$
$$\hat{J}_z = (\hat{n}_l - \hat{n}_r)/2 = \hat{n}$$

These equations connect the mean spin length $\langle \hat{J}_x \rangle$ (assuming the spin polarization to point in J_x direction) with the *coherence* $\langle \cos(\hat{\varphi}) \rangle$:

$$\langle \hat{J}_x \rangle = \frac{N}{2} \langle \cos(\hat{\varphi}) \rangle \qquad (3.12)$$

The symmetric two-mode model can be easily solved numerically by exact diagonalization up to occupation numbers $N = \mathcal{O}(10^3)$. The theoretical predictions throughout this thesis are obtained using this numerical method.

Quantum Fluctuations in Rabi, Josephson and Fock Regime

The operators $\hat{J}_i$ belong to a $J = N/2$ spin algebra. Normalizing the spin length to $j = 1/2$ in order to compare interaction and tunneling on the single particle level the Hamiltonian is $H_{norm} = N^2 E_C j_z^2/2 - 2E_J j_x$ and the essential parameter identifying the different regimes of the Josephson junction becomes obvious [35]:

$$\Lambda = \frac{N^2 E_C}{4E_J} = \frac{N\chi}{\Omega} \tag{3.13}$$

Three regimes are identified [32], the

- Rabi regime with $\Lambda \ll 1$
- Josephson regime with $1 \ll \Lambda \ll N^2$
- Fock regime with $\Lambda \gg N^2$

Here we focus on the quantum fluctuations of the ground state of the Josephson Hamiltonian in the different regimes assuming no bias ΔE and we use the spin language to describe the fluctuations.

Deep in the *Rabi regime* the ground state of Eq. (3.10) is a coherent spin state on the equator of the Bloch sphere $|\theta = \pi/2, \varphi = 0\rangle$ featuring equal fluctuations in the two orthogonal spin directions $\Delta \hat{J}_z^2 = \Delta \hat{J}_y^2 = N/4$.

With decreasing Josephson energy E_J the system enters the *Josephson regime* and quantum fluctuations $\Delta \hat{J}_z^2$ decrease at the cost of fluctuations in $\Delta \hat{J}_y^2$, however the mean spin length $\langle \hat{J}_x \rangle$ is still close to $N/2$. In the Rabi and Josephson regime the Josephson Hamiltonian (3.9) can be used to calculate the variances in the two spin directions by means of a simple analogy: The ground state features rather small fluctuations in n and φ, such that Eq. (3.9) can be expanded and one obtains a harmonic oscillator type Hamiltonian:

$$\hat{H} = E_J \frac{\hat{\varphi}^2}{2} + \left(E_C + \frac{4E_J}{N^2}\right) \frac{\hat{n}^2}{2} \tag{3.14}$$

By direct comparison to the well known harmonic oscillator result [36], the fluctuations in $\hat{J}_z = \hat{n}$ and $\hat{J}_y \approx N\hat{\varphi}/2$ are found to

$$\Delta \hat{J}_z^2 = \frac{1}{2}\sqrt{\frac{E_J}{E_C + 4E_J/N^2}} \tag{3.15}$$

$$\Delta \hat{J}_y^2 = \frac{N^2}{4} \frac{1}{2}\sqrt{\frac{E_C + 4E_J/N^2}{E_J}} \tag{3.16}$$

In the *Fock regime* the harmonic approximation breaks down since the quantum state spreads around the full Bloch sphere, resulting in vanishing coherence $\langle \cos(\varphi) \rangle \approx 0$ and mean spin length. Spin fluctuations in J_z direction (number fluctuations) are highly suppressed in the Fock regime and the remaining fluctuations correspond to less than one atom $\Delta \hat{J}_z^2 \lesssim 1$ (see Fig. 3.4).

Coherent Spin Squeezing and the Josephson Ground State

In the Rabi regime ($E_C = 0$) the ground state—a coherent spin state—is a minimum uncertainty state. Up to a small correction this remains valid over the whole range

(a)

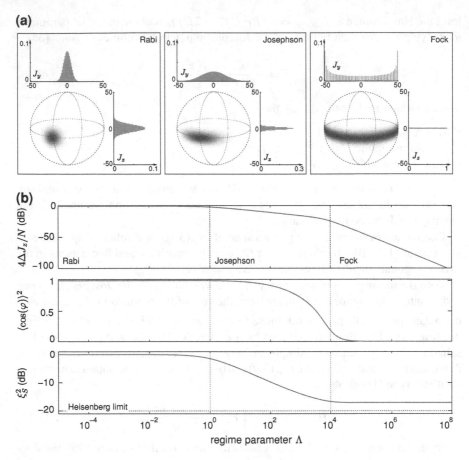

Fig. 3.4 Spin fluctuations in different regimes of the Josephson junction. **a** shows the representation of the ground state of the Josephson junction on the Bloch sphere in the three different regimes. The histograms represent the probability distribution of the quantum state over the J_z and J_y Dicke eigenbasis respectively. The distribution in J_z narrows while the variance in J_y increases. Note that the distribution of J_y in the Fock regime shows fringes with a period of $1/N$ which in principle can be used for increased sensitivity in interferometry [19, 37]. **b** shows the normalized fluctuations in J_z, the coherence $\langle \cos(\varphi) \rangle$ which is connected to the spread around the Bloch sphere and the coherent spin squeezing parameter ξ_S^2 as a function of Λ. The calculation was done for 100 atoms

of Λ [38]. The ground state shows almost minimal allowed fluctuations in $\Delta \hat{J}_z^2$ for a given coherence $\langle \hat{J}_x \rangle$ and coherent spin squeezing close to the Heisenberg limit is possible.[8] The best spin squeezing in the different regimes characterized by Λ is shown in Fig. 3.4b [39].

---

[8] The authors of reference [39] show that the Heisenberg limit can be reached within a factor of two.

3.2 Ultracold is Not Enough: Finite Temperature Effects

Temperature in Comparison to Other Relevant Energy Scales

For a dilute Bose gas confined in a three dimensional harmonic trap with mean trapping frequency $\omega_T = (\omega_x \omega_y \omega_z)^{1/3}$ the critical temperature is $T_C \approx 0.94 \, \hbar \omega_T N^{1/3}$ [9]. The chemical potential in Thomas-Fermi approximation is $\mu = \hbar \omega_T (15 N a / a_{ho})^{2/5} / 2$ where $a_{ho} = \sqrt{\hbar / m \omega_T}$ is the mean harmonic oscillator length. For our experimental parameters these two numbers are $T_C \approx 150$ nK and $\mu \approx 50$ nK.

It is hard to achieve temperatures much below the chemical potential μ by standard evaporative cooling schemes, since evaporation below μ means to "cut into the condensate". In our setup we measured temperatures down to 10 nK, a fifth of the chemical potential [40]. The fraction of non-condensed atoms can be estimated from Eq. (3.1) and is in the order of 10^{-3} meaning less than ten atoms are not in the condensate. Effects due to these few thermal atoms are negligible for our level of experimental precision. The lowest energy of transversal excitation is set by the trap frequencies ($\omega_x = \omega_y = 2\pi \cdot 425$ Hz) and it is approximately twice as large as the temperature. Exclusively the lowest many-body modes of the Josephson junction have energies much below T as discussed in the following section.

3.2.1 Collective Mode Spectrum of the Josephson Hamiltonian

The energy spectrum of the Josephson Hamiltonian for different regime parameters Λ is shown in Fig. 3.5. Our experiments are done in the Josephson regime (see Fig. 3.6b for the accessible range of Λ depending on the barrier height V_0) where for small eigenenergies ($E_k \ll 2E_J$) the many-body mode spectrum is a linear Harmonic oscillator spectrum. For these states phase and number fluctuations are rather small. The oscillator's angular frequency ω_{pl}, called *plasma frequency*, can be easily found by a taylor expansion up to second order of Eq. (3.9) and a comparison to a standard harmonic oscillator:

$$\omega_{pl} = \sqrt{E_J \left(E_C + \frac{4E_J}{N^2} \right)} \tag{3.17}$$

The equation can be reformulated as $\omega_{pl} = \sqrt{E_C E_J (1 + \Lambda^{-1})} = \Omega \sqrt{1 + \Lambda}$. In the first form one recognizes the limit in the Josephson regime $\omega_{pl} = \sqrt{E_C E_J}$ where $\Lambda \gg 1$, while the latter shows, that the Plasma frequency approaches the Rabi frequency in the Rabi regime where $\Lambda \ll 1$.

The high lying eigenstates ($E_k \gg 2E_J$) are grouped in degenerate pairs and their spacing grows quadratically with the eigenstate label $E_{k+2} - E_k = k^2 E_C / 8$, since the energy is governed by the quadratic part of the Hamiltonian [25].

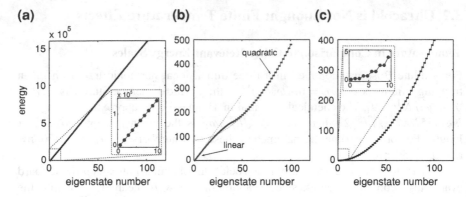

Fig. 3.5 Many-body mode spectrum of the Josephson Hamiltonian. The figure shows the energy spectrum of the many-body modes in the Rabi (**a**), Josephson (**b**) and in the Fock regime (**c**). In the Rabi regime the spectrum is completely harmonic in contrast to the Fock regime, where the eigenstates are grouped in degenerate pairs and the energy splitting between two distinct pairs grows quadratically. In the Josephson regime both features are contained in the spectrum. The lowest modes show a linear behaviour, while the high lying modes are pairwise degenerate and the spectrum is quadratic

Thermal population of the collective modes

For low but finite temperatures[9] $T \ll 2E_J$ in thermal equilibrium only the linear part of the spectrum is thermally populated and the diagonal elements of the density matrix ρ expressed in the eigenbasis of the Josephson Hamiltonian are given by

$$\rho_{kk} = Ce^{-T/k \cdot \omega_{pl}} \tag{3.18}$$

where C normalizes the density matrix. As mentioned in Sect. 3.1.2 the largest possible plasma frequency for a double-well trap realized by splitting a harmonic trap in z-direction is the longitudinal trap frequency ω_z. The ratio of temperature and longitudinal trap frequency is $T/\omega_z \approx 3$ for our parameters meaning at least the three lowest many-body states are populated at the e^{-1} level. Figure 3.6a shows the plasma frequency of our experimentally realized double-well trap with 1,600 atoms in total.

The ground state of the Josephson junction is close to a minimal uncertainty state, but finite entropy in the system, i.e. more than one populated many-body mode, causes increased fluctuations in the atom number difference n and the relative phase φ, or in other words the variances $\Delta \hat{J}_z^2$ and $\Delta \hat{J}_y^2$ in the spin directions orthogonal to the mean spin increase [40, 41]. Figure 3.7a shows the dependence of number squeezing, coherence and coherent spin squeezing ξ_s^2 as a function of temperature in the Josephson regime for $\omega_{pl} = 2\pi \cdot 60$ Hz (and $\Lambda = 150$), the largest (smallest) value reachable for our parameters. Figure 3.7b and c show these quantities for a fixed temperature and for fixed entropy (three thermally populated many-body modes)

---

[9]   In the experiment we typically have $T/2E_J, max \approx 10^{-2}$.

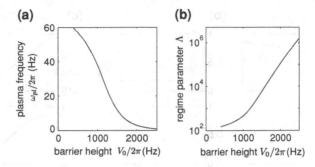

Fig. 3.6 Accessible range of plasma frequency and regime parameter. We control the Josephson junction by changing the barrier height V_0. This figure shows the range in which the plasma frequency ω_{pl} (**a**) and the regime parameter Λ (**b**) can be tuned for our double-well experiment loaded with 1,600 atoms

but different values of Λ. In the isothermal case calculated for $T = 20$ nK no spin squeezing develops—the number fluctuations are constant while the coherence vanishes with rising Λ. This is a big difference to the adiabatic case where the coherence also drops to zero but before number squeezing develops such that the coherent spin squeezed regime $\xi_s^2 < 0$ dB can be reached.

3.2.2 Strategies for Optimum Coherent Spin Squeezing

In our experiment the Bose–Einstein condensate is in a thermal equilibrium state with $T \approx 10$–20 nK right after evaporative cooling. The challenge is to achieve the best possible coherent spin squeezing given these temperature constraints. The control parameter available is the barrier height V_0 that can be dynamically changed within each experimental realization.

As discussed above an isothermal approach—condensation in a trap with fixed barrier height V_0—would not produce coherent spin squeezing since the plasma frequency ω_{pl} decreases with increasing barrier height V_0 leading to a larger number of thermally occupied many-body states. The quantum state follows the isothermal lines shown in Fig. 3.8. A better approach is to follow the adiabatic lines in Fig. 3.8. Experimentally this can be done by condensation in a small barrier height situation and subsequent adiabatic ramp up of the optical lattice V_0. In this case only a small number of many-body modes is populated initially. Since the entropy in the system stays constant, but the energy of all many-body states decreases with rising Λ, number squeezing and coherent spin squeezing develops as a result of *adiabatic cooling*.

Adiabatic Cooling and its Limits

Ramp up of the potential barrier V_0 results in a changing energy spectrum of the system. Starting in the Josephson regime with rather low Λ and high Josephson

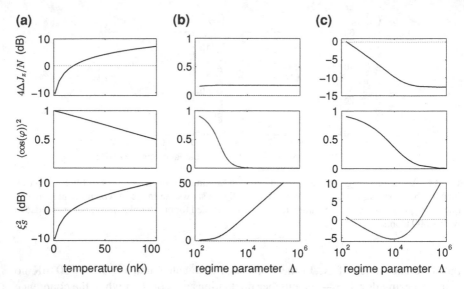

Fig. 3.7 Spin squeezing at finite temperature. From *top* to *bottom* the evolution of number squeezing, coherence and coherent spin squeezing are shown. In **a** these observables are given versus temperature for $\Lambda = 150$ and $\omega_{pl} \approx 60$ Hz. Only for temperatures below approximately 15 nK number squeezing and coherent spin squeezing $\xi_S^2 < 0$ dB are present. **b** The isothermal evolution of the quantum state for $T = 20$ nK is shown versus regime parameter Λ. Almost constant number fluctuations and degrading coherence prevent coherent spin squeezing to develop. Part **c** shows the results of an adiabatic calculation assuming initially three thermally populated many-body modes. For intermediate values of Λ coherent spin squeezing can be reached. The calculations were done for values close to our experimental parameters, especially $E_C = 2\pi \cdot 1$ Hz and $N = 1,600$. E_J was controlled by changing Λ. For the connection between Λ, the plasma frequency ω_{pl} and the barrier height V_0 in our experiment see Fig. 3.6

energy E_J, the Josephson energy decreases which has two major implications: The plasma frequency decreases leading to adiabatic cooling, but the boundary region between the linear and quadratic part of the spectrum also moves towards the lower eigenstates. Up to intermediate barrier heights V_0 where only a negligible fraction of the occupied states lie in non-harmonic part of the spectrum the quantum state of the system can be described by the harmonic oscillator Hamiltonian given in Eq. (3.14). For a given thermal density matrix ρ the fluctuations in the atom number difference $\hat{n}$ are given by

$$\langle \hat{n}^2 \rangle = \text{Tr}(\rho \hat{n}^2) = \frac{\sum_k e^{-E_k/T} \langle k|\hat{n}^2|k \rangle}{\sum_k e^{-E_k/T}} \tag{3.19}$$

where the eigenenergies are $E_k = k \cdot \omega_{pl}$ and the matrix element for harmonic oscillator eigenfunctions $|k\rangle$ is $\langle k|\hat{n}^2|k \rangle = (k + 1/2)\sqrt{E_J/E_C}$. It follows that

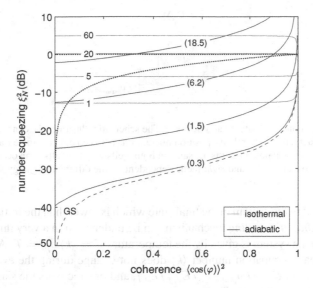

Fig. 3.8 Phase diagram at finite temperature. The phase diagram—number squeezing versus coherence—is shown for the same parameters that have been used in Fig. 3.7. The *dotted black lines* illustrate the limits for number squeezing and coherent spin squeezing. The *dashed line* corresponds to the ground state of the Josephson junction. *Gray solid lines* are isothermals at different temperatures indicated on the *left*. The *black lines* correspond to an adiabatic evolution of the system where the number of initially thermal populated many-body modes is given in brackets. The regime parameter Λ—or experimentally the barrier height V_0—increases for each line from the *right* to the *left*

$$\langle \hat{n}^2 \rangle = \sqrt{\frac{E_J}{E_C}} \left(\frac{3}{2} + \frac{1}{e^{\omega_{pl}/T} - 1} \right) \approx \frac{T}{E_C} \qquad (3.20)$$

where $\omega_{pl}/T \ll 1$ for the last approximation.[10]

Assuming adiabatic evolution, the initial ratio of $\omega_{pl,i}/T$ stays constant, but the matrix element is evaluated at the final value of the Josephson energy $E_{J,f}$.[11] This results in decreased number fluctuations as compared to the initial state:

$$\langle \hat{n}^2 \rangle^{(f)} \approx \frac{T}{E_C} \sqrt{\frac{E_{J,f}}{E_{J,i}}} = \frac{T_{\text{eff}}}{E_C} \qquad (3.21)$$

The argumentation holds as long as the distribution stays thermal, i.e. only the linear part of the spectrum is populated. An effective temperature $T_{\text{eff}} = T\sqrt{E_{J,f}/E_{J,i}}$

---

[10] $\hat{n}$ and $\hat{\varphi}$ are symmetric variables in Eq. (3.14). The fluctuations in the relative phase $\Delta\hat{\varphi}^2 \approx T/E_J$ are obtained in the same way as described for $\Delta\hat{n}^2 = \langle\hat{n}^2\rangle$, but replacing the matrix element by $\langle k|\hat{\varphi}^2|k\rangle = (k + 1/2)\sqrt{E_C/E_J}$.

[11] For our parameters $E_{C,f} \approx E_{C,i} = E_C$ holds.

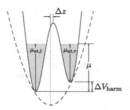

Fig. 3.9 Position noise translates to atomic noise. The schematic illustrates the connection of number difference fluctuations and relative position fluctuations between optical dipole trap and optical lattice Δz. The chemical potential μ is the same in both wells which results in an occupation number difference between the *left* and *right* mode dependent on the differential energy shift ΔV_{harm} between them

can be assigned to the system in the final state which is lower than the initial temperature T. This *adiabatic cooling* mechanism can be understood in a very intuitive way: The energy in the system is given by the temperature $E = \langle k \rangle \omega_{pl} = T$. Adiabaticity means the mean occupation number $\langle k \rangle$ does not change during the evolution and therefore $T_f / T_i = \omega_{pl,f} / \omega_{pl,i} = \sqrt{E_{J,f} / E_{J,i}}$, and one recovers the same result as above for $T = T_i$ and $T_{eff} = T_f$.

Adiabatic cooling reaches its limit in the Fock regime, when all states are double degenerate and the energy splitting between the pairs increases quadratically. The matrix element $\langle k | \hat{n}^2 | k \rangle = k^2$ has to be calculated for $|k\rangle$ being eigenstates of the number difference operator and the degeneracy has to be taken into account. The cooling limit follows from

$$\langle n^2 \rangle = \frac{2 \sum_k k^2 e^{-2k\omega_{pl}/T}}{\sum_k e^{-E_k/T}} \approx \frac{1}{2} \left(\frac{T_i}{\omega_{pl,i}} \right)^2 \tag{3.22}$$

and it is determined by the initial entropy which is measured by $T_i / \omega_{pl,i}$.

3.3 Quantum Fluctuations in Few-Well Potentials: Experimental Challenges

3.3.1 Position Stability of the External Trapping Potentials

As discussed above finite temperature limits the minimum achievable fluctuations $\Delta \hat{J}_z^2$ and $\Delta \hat{J}_y^2$ and therefore the coherence $\langle \cos(\varphi) \rangle$. Beside cooling to ultra low temperatures the second technical challenge is the position stability of the different optical dipole traps [27]. Relative movement of the dipole trap with respect to the position of the potential barrier causes fluctuations of the atom number difference n between the left and the right well. Figure 3.9 illustrates this situation. The optical dipole trap generates harmonic confinement in longitudinal direction $V_{harm} = m\omega_z^2 z^2 / 2$, where m is the atomic mass. Fluctuations of the energy difference ΔV_{harm} between the two potential minima separated by d due to fluctuations of the barrier position Δz is given by:

$$\Delta V_{\text{harm}} = \frac{\partial V_{\text{harm}}}{\partial z} \Delta z = m\omega_z^2 d \Delta z \qquad (3.23)$$

In the local density approximation [10] the overall chemical potential is $\mu = V_{\text{harm}}(z) + \mu_{at}(z)$ and in equilibrium it is the same for the two wells. The contribution due to interatomic interactions $\mu_{at}(z)$ balances the change in $V_{\text{harm}}(z)$: $\Delta V_{\text{harm}} = \Delta \mu_{at}$

$$\Delta \mu_{at} = \frac{\partial \mu_{at}}{\partial n_l} \Delta n_l = E_C \Delta n \qquad (3.24)$$

With the experimental parameters for the double-well potential $E_C \approx 2\pi \cdot 1$ Hz and $\omega_z \approx 2\pi \cdot 60$ Hz we obtain

$$\Delta z = \frac{E_C}{m\omega_z^2 d} \Delta n \lesssim 125\,\text{nm} \qquad (3.25)$$

as the position fluctuation leading to an extra noise of the same order as the shot noise limit $\xi_N^2 = 0$ dB for 2,000 atoms. Since these technical induced fluctuations add to the variance caused by the atomic quantum state, their magnitude has to be much smaller than the shot noise level in order to measure a reasonable amount of number squeezing. As a figure of merit, position fluctuations of 60 nm between different experimental realizations limit the best observable number squeezing to $\xi_N^2 \approx -6$ dB.

Ultra Stable Optical Traps

Figure 3.10a shows schematically the setup of the laser beams necessary to generate the double-well potential. As as described in Sect. 3.1.2 and in Ref. [23] the setup consists of four laser beams of which one generates the main dipole trap, an additional dipole trap beam increases the longitudinal confinement[12] and two beams interfere to generate the one dimensional lattice with lattice spacing $d = 5.7\,\mu$m. The dipole trap potential minimum is positioned such that it coincides with a node of the red detuned optical lattice, thus making up the double-well potential. The position of the interference pattern is actively stabilized at a reference position which is chosen as close as possible to the atomic cloud, but outside the vacuum chamber. Position feedback is implemented by control over the relative phase between the two lattice beams [27]. However the positions where the interference pattern is interferometrically stabilized and where the optical trapping beams for the dipole traps are launched are macroscopically spaced by approximately 20 cm. Relative position stability in the order of a few tens of nanometers is therefore a technical challenge. We mount all optical beams—avoiding mechanical stress as good as possible—on a

---

[12] The frequency of this beam differs from the frequency of the main dipole beam by 30 MHz to average their interference pattern.

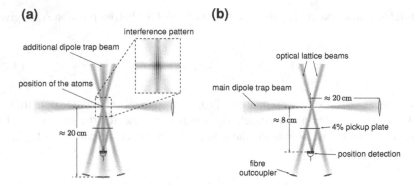

Fig. 3.10 Laser beam configuration for the optical potential. **a** illustrates the beam setup required to generate a double-well trap at the crossing point of the four laser beams. In part **b** one of the dipole trap beams is omitted which results in a smaller longitudinal trapping frequency and decreased sensitivity of the atomic fluctuations to position fluctuations. Smaller longitudinal confinement results in more than two populated lattice sites. For the chosen total atom number of 5,300 atoms in this trap typically six to seven wells are populated. The important distances between the main elements are indicated and the interferometric position stabilization of the optical lattice is schematically shown

massive casted block of AlMg4,5Mn aluminum alloy for optimal passive stability. This block is hold to the optical table by its own weight and it is carried by three steel balls similar to a standard mirror mount design. The required stability of ca. 60 nm is still hard to achieve day to day[13] but the experimental results presented below suggest that we are close to this level of stability.

3.3.2 From Two to Few: The Six-Well Trap

Quantitative measurements of number and phase fluctuations require long measurement time since the repetition rate of our experiment is one minute and we need approximately 100 experimental repetitions per parameter set to have reasonable statistics. Fulfilling the double-well stability requirements on a timescale of a few hours is experimentally hard to achieve.

In order to decrease the sensitivity to position fluctuations we omit the trapping beam that provides extra longitudinal confinement (Fig. 3.10b). The longitudinal frequency of the trapping potential is in this case $\omega_z = 2\pi \cdot 20$ Hz for $\omega_x = \omega_y = 2\pi \cdot 425$ Hz. The spatial stability requirement given in Eq. (3.25) is proportional to the inverse quadratic trap frequency ω_z^{-2}. A reduction of ω_z by a factor of three from the double- to the few-well geometry relaxes the required position stability to $\Delta z \approx 350$ nm for $\xi_N^2 \approx -10$ dB maximum number squeezing. Loading a ^{87} Rubidium Bose–Einstein condensate of approximately 5,300 atoms into this trap results in a chemical

---

[13] The thermal expansion coefficient of alluminium is ca. 23×10^{-6} /K at room temperature, leading to a temperature stability requirement of 10 mK over a few hours, the typical duration of the experiment.

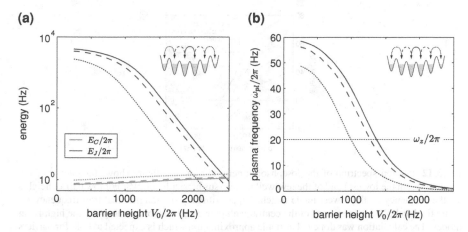

Fig. 3.11 Local parameters of the Josephson junction array. The local Charging and Josephson energy for each Josephson junction in the few-well trap is calculated following the recipes given in Sect. 3.2.1. We choose a geometry with two equally populated wells in the center as shown in the insets and the calculation was done for 5,300 atoms with the trapping parameters as given in the main text. The line style distinguishes the different well pairs. In **a** the Charging and Josephson energy is plotted, while **b** shows the resulting local plasma frequency. Interestingly the maximum local plasma frequency is much larger than the longitudinal trap frequency ω_z

potential of $\mu \approx 1$ kHz and the condensate has a longitudinal extension of ca. 40μ m. Six to seven lattice sites are populated after superposing the one dimensional optical lattice where the actual number of sites depends on the relative position of lattice and dipole trap potential minimum. Our detection system allows for the measurement of the atom number in each of the wells and for the measurement of the relative phase between two next neighbors (see Sect. 3.4). This provides access to the local spin variables of two neighboring wells and we approximate each well pair as a single Josephson junction. Figure 3.13a, b shows absorption pictures taken in the double- and few-well trap situation respectively.

Temperature in the Few-Well Case

The few-well configuration can be described as an array of non-identical Josephson junctions. The mean field Gross–Pitaevskii wavefunction is used to calculate the Charging energy $E_C^{(m)}$ and Josephson energy $E_J^{(m)}$ for each junction m as given in Eq. (3.6) and (3.7). In Fig. 3.11 the results for our experimental parameters are plotted versus barrier height V_0.

In order to estimate the effect of finite temperature in the double-well case the argument $\omega_{pl} \to \omega_z$ for vanishing barrier height $V_0 \to 0$ was used in Sect. 3.2. Figure 3.11b shows that this argument does not hold any more in the few-well case. The extension of one local Josephson junction is much smaller than the extension of the condensate in longitudinal direction. Therefore the local plasma mode corresponds to a rather short wavelength—high energy—excitation as compared to the trap dipole mode. This argumentation is a simplification since the energy spectrum in the few-well situation shows a band structure with $M - 1$ modes per band for M wells.

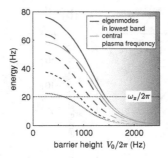

Fig. 3.12 Phonon spectrum of the Josephson junction array. The figure shows the energy of the eigenmodes in the lowest band of the few-well trap versus barrier height V_0. For low barrier heights V_0 the frequency of the lowest mode matches approximately the longitudinal trap frequency ω_z while the local plasma frequency of the central well pair compares to the energy of the high lying modes. The calculation was done in harmonic approximation which is expected to hold for barriers up to $V_0 \approx 1.5$ kHz (illustrated by the shading). For greater barriers the local Josephson energies are comparable to the Charging energies and the system is close to the Fock regime (see Fig. 3.11)

Figure 3.12 shows the eigenmodes in the lowest band of our few-well trap for 5,300 atoms. The calculation was done in harmonic—phonon—approximation treating the system as coupled oscillators with different masses ($E_C^{(m)}$) and spring constants ($E_J^{(m)}$). Exact numerical diagonalization following reference [42] reveals the eigenmodes. The harmonic approximation does certainly not hold any more in the high lattice case where each Josephson junction enters the Fock regime. However for the qualitative arguments presented here only the low lattice regime is important.

The local plasma mode is not an eigenmode of the problem, but its energy is in the upper part of the first band. We expect that a few of the eigenmodes overlap with the local plasma modes but as argued above the short wavelength modes should contribute most. The discussion here shows that the local treatment is an approximation neglecting the long wavelength excitations in the system.

Comparing the frequency of the plasma excitation of the central well pair with the typical temperature, we obtain $T/\omega_{pl} \approx 3$, approximately the same number that was found in the double-well situation. Our experiments in the double- and few-well situation are effectively in a similar entropy regime. A more sophisticated theoretical treatment of our few-well system can be found in [43].

3.4 Spin Squeezing Across a Josephson Junction: Experiments

3.4.1 Detection of Number Difference and Relative Phase

In order to measure coherent spin squeezing $\xi_s^2 = N \Delta \hat{J}_z^2 / \langle \hat{J}_x \rangle^2$ fluctuations in one spin direction $\Delta \hat{J}_z$, the mean spin length $\langle \hat{J}_x \rangle$ and the total number of atoms N have to be detected. Measurement of the atom number difference $n = J_z$, its fluctuations, and

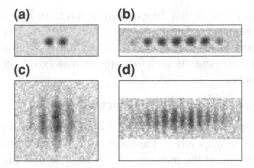

Fig. 3.13 Typical single shot pictures showing number and phase detection. The *left* column of the figure shows number (**a**) and phase (**c**) detection for the double-well case. Both variables can also be locally measured in the few-well situation. Number detection (**b**) is straightforward while for the phase measurement (**d**) the correct expansion time has to be chosen in order to allow only condensates from neighboring lattice sites to interfere (see also Fig. 3.15)

the total atom number N is straight forward and the only requirement is an accurately calibrated, linear imaging system with single lattice site resolution. Measurement of the mean spin length $\langle \hat{J}_x \rangle = N \langle \cos(\varphi) \rangle /2$ is possible via the measurement of the relative phase φ which is revealed from an interference pattern between the wavefunctions from two neighboring wells. A detailed discussion of the statistical analysis to calculate atom number fluctuations and the coherence follows later in this section, but in the next paragraph we discuss the requirements on the detection system to measure n and φ in a single realization of the experiment.

We installed an absorption imaging system with single lattice site resolution that was developed in our group. Details on the setup, its calibration and on the imaging sequence can be found in appendix A. Figure 3.13a, b shows images of the condensate in the double-well and few-well trap where the imaging parameters were adjusted for on site atom number detection. The images in Fig. 3.13c, d correspond to relative phase measurements and show single shot interference patterns for the two and few-well case. In the few-well situation the relative phase between two neighboring wells is deduced from local interference patterns. Therefore the detection requires the correct expansion time of the condensates such that only the wavefunctions from two next neighboring lattice sites overlap. Fringes are observed after a short expansion in the harmonic trap in absence of the lattice potential (2 ms) followed by a free expansion (400–900 μs). In order to choose the proper timing, we image the cloud after different free expansion times and observe the formation of the interference pattern. For too short expansion times, clouds released from neighboring wells do not overlap which is easily seen in the images. In the case of a low lattice depth, all wells are in phase leading to a maximum of the interference pattern at the middle positions between the wells. We choose the timing such that this central maximum is clearly visible.

Number squeezing and coherence measurements require to measure the statistical quantities Δn^2 and $\langle \cos(\varphi) \rangle$. However the detection process of the Bose–Einstein condensate is destructive such that the experiment has to be repeated with the same parameters in order to measure distributions of n and φ. Typically one dataset consists of 25–40 experimental "shots" and in order to reduce statistical uncertainties we average three to four of these datasets such that approximately 100 single experimental realizations contribute.

Occupation Number Difference and its Fluctuations

The atom number in each well is extracted from an absorption picture where the signal on each pixel corresponds to the column density of atoms convolved with the point spread function of the imaging system ($\sigma_{\mathrm{psf}} \approx 700$ nm). We choose the number of atoms in the trap and the expansion time before imaging such that the detected signal can be unambiguously assigned to individual wells (see Appendix A for more details). By pixel-wise summation over the area on the picture containing more than 99% of the atoms per well we obtain the occupation number. We checked that the detected total atom number and its fluctuations do not depend critically on the size of the summation area. Given the atom number in each well the calculation of the atom number difference for each well pair $n = (n_l - n_r)/2$ is straight forward but the deduction of the variance $\Delta \hat{n}^2$ of the quantum state requires some caution.

A Grubb outlier detection algorithm [44] is used to filter the atom number difference n for rare outliers caused by technical problems. It detects typically zero but maximally 1–2 points per dataset (at a 5% significance level). Due to possible slow drifts of the trapping potentials (on the timescale of one hour) we correct each dataset by removing a linear slope. Statistical simulations were performed to test this procedure and biasing was found to be negligible.

For each dataset, we define $p = \langle n_l/N \rangle$ the probability for an atom to be found in the left well where $N = n_l + n_r$ is the total atom number per wellpair. If $p \neq 1/2$ the atom number difference n depends on the total atom number as $n = (p - 1/2)N$. Therefore fluctuations in the total atom number between different experimental runs contribute to the measured variance. We compute

$$\Delta n_{raw}^2 = \langle [(n_l - n_r)/2 - (p - 1/2)N]^2 \rangle \tag{3.26}$$

in order to avoid taking these fluctuations into account. Since p is typically close to 1/2 this correction has only a small effect.

Additional noise $\Delta n_{l(r),\mathrm{ps}}^2$ in the atom number $n_{l(r)}$ per well due to photon shot-noise from the detection process contributes to the variance Δn_{raw}^2. We deduce this extra noise as the sum over the variance per CCD pixel in the integration area where the contribution per pixel is inferred from the light intensity on the absorption and on the reference picture. A measured CCD camera noise calibration curve relates the mean counts to the variance per pixel and permits to calculate the additional atomic variances $\delta n_{l(r),\mathrm{psn}}^2$ for each experimental realization (see Appendix A). We subtract this contribution

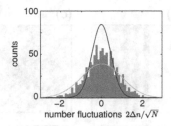

Fig. 3.14 Squeezed distribution of the atom number difference. Shown are the detected atom number difference fluctuations for a number squeezed state where in total approximately 1,000 measurements contribute. The histogram shows the experimental raw data after filtering by the Grubb outlier test and linear drift removal. The *gray curve* is the expected distribution for a shot noise limited quantum state without experimental noise, while the *black curve* is the inferred distribution (assuming a gaussian shape) after subtracting known noise from the experimental data

$$\Delta n_{psn}^2 = [1/4 + (p - 1/2)^2]\langle \delta n_{l,psn}^2 + \delta n_{r,psn}^2 \rangle \qquad (3.27)$$

and obtain the corrected number fluctuations:

$$\Delta n^2 = \Delta n_{raw}^2 - \Delta n_{psn}^2 \qquad (3.28)$$

Number squeezing is detected when the measured fluctuations Δn^2 are lower than expected for a binomial distribution—the shot noise limit.

$$\xi_N^2 = \frac{\Delta n^2}{p(1 - p)\langle N \rangle} < 1 \qquad (3.29)$$

Figure 3.14 shows the histogram of measured atom number differences where several datasets were combined resulting in approximately 1,000 total counts. The two gaussian curves represent the distributions expected for a shot noise limited state and for the detected number squeezed state where the measured variance is corrected for total atom number fluctuations and photon shot noise.

Relative Phase and Coherence

We measure the coherence of the quantum state for the same experimental parameters as chosen for the corresponding number fluctuation measurement. In order to deduce the relative phase between two neighboring wells we Fourier transform the transversally averaged interference pattern and extract the phase φ from the dominant frequency component. In the few-well situation we slice the picture at the center position of each well and infer the relative phase between two wells based on these slices (see Fig. 3.15). We calculate the coherence $\langle\cos(\varphi)\rangle$ by ensemble averaging $\cos(\varphi)$ over each dataset where φ is obtained from the single shot interference patterns. Figure 3.15 shows profiles of exemplary datasets for high and low coherence in the double- and few-well system. Each pixel row of each picture is a vertically averaged single shot profile where the depth of gray shading corresponds

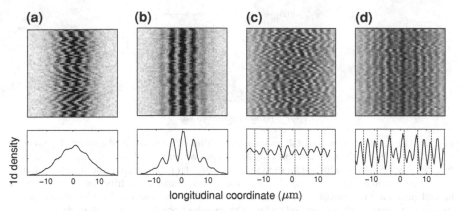

Fig. 3.15 Phase coherence for different barrier heights. **a** and **b** show phase measurements in the double-well situation for a high barrier in **a** and for a low barrier in **b**. The two columns **c**, **d** on the *right* show the respective phase measurements in the few-well situation. Each line in the images corresponds to an transversally averaged profile of a single experimental realization. The graphs below show the average of the pictures above in vertical direction. In the few-well case the boundaries between different well pairs are indicated. For a high barrier V_0 (**a** and **c**) the relative phase measured in each shot is almost random and the average visibility is decreased, while a stable relative phase is observed for low barriers (**b** and **d**)

to the number of atoms per pixel. The ensemble averaged profile is shown below, which also reveals the coherence from its visibility.

In the few-well situation each slice contains only 11 pixels, therefore photon shot noise in the individual images is one limiting factor for the phase estimation precision. Experimentally we find smallest fluctuations when calculating the phase of the interference fringe relative to the absolute position of the fringe on the camera sensor. However from repeated measurements of the center of mass position of a small atomic cloud we extract root mean square fluctuations of the imaging systems position in the order of one pixel. Both effects, finite signal to noise and position fluctuations, limit the phase precision to $\Delta\varphi_{min} \approx 23°$ in the few-well situation. We do not correct for the this additional noise leading to a systematic underestimation of the true coherence across the junction.

It is more general to extract the coherence from the ensemble averaged interference patterns shown in the lower part of Fig. 3.15 than ensemble averaging the individual phases, since in a non-two-mode situation the single shot visibility of the patterns might be already decreased. The drawback of this method however is an underestimation of the coherence due to the finite resolution of the imaging system. We found that the single shot visibility is compatible with unity taking the finite resolution of the imaging system into account (Fig. 3.16), which justifies our method to compute the coherence.

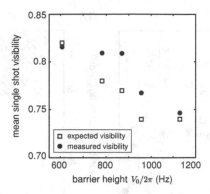

Fig. 3.16 Single shot visibility of the interference patterns. The fringe visibility observed in a single experimental realization is compatible with unity when the finite optical resolution is taken into account. The figure shows the observed average single shot visibility (*solid circles*) compared to the expected visibility for a pure two-mode situation including the effect of finite optical resolution (*open squares*). Experiment and prediction agree within a few percent. The trend in the observed data which is due to the dependence of the visibility from the fringe wavelength is reproduced by the theory. The wavelength decreases with rising barrier since the in the tighter trapping results in increased kinetic energy at the time of release

3.4.2 Measuring the Timescale for Adiabatic Changes

In Sect. 3.2 we discussed the effects of finite temperature on the fluctuations in atom number difference and relative phase. Adiabatic cooling was presented as one approach to achieve number squeezing despite of thermally induced fluctuations. In order to change the state of the system adiabatically the timescale τ in which the potential barrier V_0 is ramped up and therefore the plasma frequency ω_{pl} is changed has to be smaller than the inverse plasma frequency itself [39, 46]. Therefore it is hard to drive the system adiabatically into the Fock regime where $E_C/4E_J \gg 1$ if the Charging energy E_C is small since this requires ramp times $\tau \ll E_C^{-1}$. For our parameters ramp times τ on of the order of a few tens of seconds are necessary to reach the ultimate cooling limit given in equation (3.22). These long times are not realizable in the experiment without significant perturbation of the system due to the environment which leads to particle loss and heating [40].

Due to these disturbing effects the experimental duration should be chosen as short as possible and the optimal ramp duration for a given situation is best found experimentally. We start with a Bose–Einstein condensate in a low barrier trap of $V_0 = 2\pi \cdot 430$ Hz. Now we ramp up the potential barrier in a linear manner to a fixed end height of $V_0 = 2\pi \cdot 1,650$ Hz. We repeat this experiment varying the total ramp time and measure the number squeezing parameter ξ_N^2. Figure 3.17 shows that ramps with a slope smaller than $2\pi \cdot 10$ Hz/ms are found to be adiabatic within the detection accuracy of our experiment. In the double-well situation with 1,100 atoms in total we measured up to very long ramp times in the order of a few seconds. Number squeezing however levels around $\xi_N^2 \approx -2$ dB which is explained taking

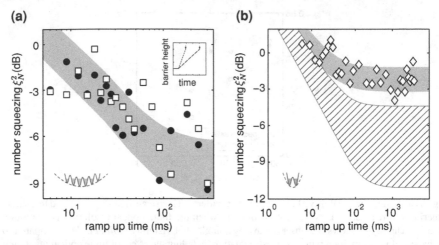

Fig. 3.17 Timescales for adiabatic barrier ramps. We ramp up the barrier height V_0 from $2\pi \cdot 430$ Hz to $2\pi \cdot 1,650$ Hz in a linear way varying the total ramp time. **a** shows the results of the measurements for the few-well situation. The gray shaded area is a theoretical simulation within the local two-mode model assuming temperatures between 20 and 40 nK. The two different symbols correspond to the two central well pairs in the optical lattice. This figure is originally published in [45]. **b** shows the same experiment but in the double-well geometry. Temperature alone as the limiting factor for number squeezing can not explain the observed data (*hatched area*). Including relative position fluctuations of dipole trap and lattice beams, the experimental observation is reproduced for the same temperatures as in the few-well case (*gray shaded area*)

position noise into account as detailed below. The data shows even an upward trend for long ramps which is attributed to heating and atom loss. We load 5,300 atoms in total into the few-well trap corresponding to a occupation number of $N \approx 2,200$ atoms in each of the two central well pairs. Number squeezing increases from $\xi_N^2 \approx -2$ dB for a low barrier to the best observed value of $\xi_N^2 = -6.6^{+0.8+0.8}_{-1.0-0.8}$ dB. This number is calculated by averaging over several datasets such that approximately 1,000 experimental realizations contribute. The given uncertainties are one standard deviation statistical errors of the mean over all datasets followed by an upper bound of 20% for systematic errors due to a possible calibration error of the atom number detection. We find adiabatic cooling of approximately a factor of three. At $V_0 = 2\pi \cdot 1,650$ Hz the system is not yet in the Fock regime (see Fig. 3.11), however the highest occupied modes are not any more in the linear part of the spectrum. This results in large fluctuations in the relative phase and a loss of coherence across the Josephson junction (see Fig. 3.15). Since the main experimental goal is to generate many-body entanglement and coherent number squeezing we tried to find the optimum ramp time to this intermediate barrier height. Furthermore the detected number squeezing of $\xi_N^2 = -6.6$ dB means fluctuations of only 10 atoms out of 2,200 which is close to the detection threshold of our imaging system.

The gray shaded area in Fig. 3.17 shows the result of a numerical simulation of the two-mode Josephson Hamiltonian using the dependence of the Hamiltonian parameters E_C and E_J on the barrier height V_0 shown in Figs. 3.11 and 3.3. In the theory the initial temperature was adjusted to fit the data and the upper bound of the area corresponds to $T = 40$ nK while the lower line corresponds to $T = 20$ nK. In order to fit the double-well data, we need to take position fluctuations of the trapping beams into account which limit the best observable number squeezing. Position fluctuations with a root mean square amplitude of $\Delta z \approx 80$ nm explain the data (see Sect. 3.3)

We restricted the change of the optical lattice intensity to linear ramps since they require only one parameter, the ramp time τ, for given initial and final barrier height. In future experiments the achieved number squeezing might be optimized further using custom ramp shapes obtained from optimal control schemes [47, 48].

3.4.3 Coherent Spin Squeezing and Many-Body Entanglement

Optimizing Coherent Spin Squeezing

Knowing the timescales for the barrier ramp the challenge is to find a final barrier height V_0 where the amount of number squeezing and phase coherence allows to achieve coherent spin squeezing $\xi_S^2 < 0$ dB.

In order to answer this question experimentally we follow a barrier ramp with a slope of $2\pi \cdot 4$ Hz/ms (few-well trap) from a low barrier situation where the condensate is obtained to a variable end value.[14] Figure 3.18a shows the results for the few-well situation. Open and solid data points correspond to the two central well pairs in the lattice each populated by approximately 2,200 atoms. The phase coherence $\langle \cos(\varphi) \rangle$ is plotted in the upper panel and below the measured number squeezing ξ_N^2 is shown. For barrier heights below ca. $2\pi \cdot 1,000$ Hz we find high phase coherence and simultaneously a considerable amount of number squeezing. Averaging the measurements between $V_0 = 2\pi \cdot 650$ Hz and $V_0 = 2\pi \cdot 900$ Hz we calculate the best coherent spin squeezing:

$$\xi_S^2 = -3.8^{+0.3+0.8}_{-0.4-0.8} \text{ dB} \tag{3.30}$$

This value is obtained from approximately 500 phase and number difference measurements and the uncertainties are one standard deviation statistical errors of the mean followed by bounds on possible systematic errors. The systematic error is due to uncertainties in the imaging calibration and due to the afore mentioned underestimation of the phase coherence.

The gray shaded areas are the predictions from a two-mode approximation of the Josephson Hamiltonian assuming adiabatic evolution, where the initial temperature was adjusted to fit the data. The lower bound corresponds to $T = 10$ nK and the upper bound to $T = 30$ nK, implying an entropy of three to ten thermally populated many-body states across the junction. Reasonable good agreement with the data confirms the adiabatic cooling model and the local two-mode approximation presented above.

---

[14] We ramp from $V_{0,i} = 2\pi \cdot 430$ Hz for all end values $V_0 \geq 2\pi \cdot 430$ Hz and from $V_{0,i} = 2\pi \cdot 250$ Hz for all other end values.

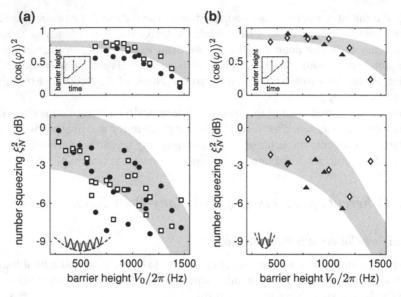

Fig. 3.18 Phase coherence and number squeezing. The experiments carried out to optimize coherent number squeezing are shown here. The *upper* row shows the evolution of the coherence, while the *lower* row shows the number squeezing. Part **a** summarizes the results of the measurements done in the few-well situation. As shown in the inset we follow a fixed slope to different barrier heights and before detection we ramp up the optical lattice within 10 ms to the end value of $2\pi \cdot 1,650$ Hz. *Solid* and *open* symbols correspond to the two central well pairs in the optical lattice populated with 2,200 atoms each. The *gray* shaded area shows the result of a local two-mode calculation assuming adiabatic evolution with three to ten populated many-body modes. Part **b** shows the equivalent measurement for the double-well with 1,600 atoms in total. *Solid* and *open* symbols correspond to two distinct measurements where the slope of the barrier ramp was different (see main text). The *gray* shaded area is the result of a two-mode calculation with the same assumptions as in the few-well case. This figure is originally published in [45]

For the phase measurement a maximal coherence of $\langle \cos \varphi \rangle^2 = 0.85$ was taken into account to match the theory with the data, limited by the phase detection method as described in Sect. 3.4.1. Since the phase measurement relies on the overlap of the wavefunctions from two neighboring wells after expansion, we ramp the barrier within 10 ms to $V_0 = 2\pi \cdot 1650$ Hz for all final barrier heights lower than this value. This is necessary since the expansion velocity after release depends on the onsite interaction energy and the intra-well trap frequency and therefore on the barrier height and the atom number.

Figure 3.18b summarizes the results for the double-well trap. The different symbols represent two different measurements where the barrier was ramped up with a slope of $2\pi \cdot 2$ Hz/ms (solid triangles) and $2\pi \cdot 8$ Hz/ms (open diamonds). As in the few-well case adiabatic evolution within the two-mode model is assumed for the same initial temperature and entropy. No upper limit for the phase coherence is necessary here since a longer expansion time allows to observe the interference pattern spread over many pixels such that the deteriorating effects of the detection noise are negligible on this level of precision. Averaging the data between $V_0 = 2\pi \cdot 650$ Hz and $V_0 = 2\pi \cdot 1,200$ Hz we measure a best coherent spin squeezing of $\xi_S^2 = -2.3^{+0.2+0.8}_{-0.6-0.5}$ dB.

Comparing the results for the few-well and double-well situation we find a better coherent spin squeezing in the few-well situation. This is explainable by the stringent spatial stability requirement in the double-well case (see Sect. 3.3) making it hard to obtain better number squeezing.

The best measured spin squeezing of $\xi_S^2 = -3.8$ dB potentially allows for a phase precision gain of $1 - \frac{\Delta\varphi_{sq}}{\Delta\varphi_{sql}} = 35\%$ in an ideal Ramsey type interferometer (see Sect. 2.4). $\Delta\varphi_{sql}$ is the phase error given by the standard quantum limit while $\Delta\varphi_{sq}$ denotes the phase error that could be obtained using a spin squeezed state for the same total atom number.

Systematic Deviations from the Theory

In the coherence graph for the few-well situation (Fig. 3.18a) a systematic overestimation of the coherence by the two-mode theory is visible, while the observed dependence of the number squeezing is reproduced. We attribute the larger phase fluctuations to contributions from the longer wavelength modes present in the few-well situation (see Sect. 3.3). Multi mode effects in the few well case as well as onsite number squeezing have been analyzed in [43].

Many-Body Entanglement

In Sect. 2.3 we discussed the connection of number squeezing and coherent spin squeezing to many-body entanglement. We briefly summarize the arguments given there: While number squeezing $\xi_N^2 < 0$ dB detects entanglement in a symmetric situation, coherent spin squeezing $\xi_S^2 < 0$ dB requires entanglement without any symmetry assumption. In Fig. 3.19 we plot our results in this context.

Data points shown are obtained by averaging the data in Fig. 3.18 and the symbols are chosen correspondingly. Solid data points are calculated averaging the measurements between $V_0 = 2\pi \cdot 650$ Hz and $V_0 = 2\pi \cdot 900$ Hz ($V_0 = 2\pi \cdot 650$ Hz and $V_0 = 2\pi \cdot 1,200$ Hz) while open symbols represent averaged data above $V_0 = 2\pi \cdot 1,300$ Hz ($V_0 = 2\pi \cdot 1,400$ Hz) for the few- (double-) well case.[15] For the high barrier situation phase coherence is lost due to temperature induced fluctuations while number squeezing increases slightly as compared to the intermediate barrier height regime (solid symbols). Assuming a symmetric two-mode situation which is valid for a Bose–Einstein condensate restricted to two modes all points shown correspond to a non-separable many-body density matrix. However the solid symbols are located below the curved dotted black line, which is the boundary for coherent spin squeezing and many-body entanglement is unambiguously detected.

---

[15] In the double-well case two more data points not shown in Fig. 3.18 between $V_0 = 2\pi \cdot 1,650$ Hz and $V_0 = 2\pi \cdot 1,800$ Hz contribute to the averaging.

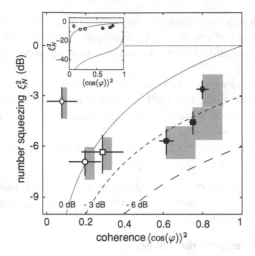

Fig. 3.19 Many-body entanglement. We summarize the results shown in Fig. 3.18 in the context of many-body entanglement. Diamonds correspond to the double-well measurements and *circles* (*squares*) to the two central well pairs in the few-well trap. One standard deviation statistical error bars are given and the *gray* shaded boxes are systematic errors. They are asymmetric in the horizontal direction since our method to measure the coherence suffers from a systematic underestimation. The phase diagram—number squeezing versus coherence—is divided in three regions: The coherent spin squeezed region below the *curved dotted line* (marked as 0 dB), the number squeezed region below the straight *dotted line* and the non-squeezed region above. For intermediate barrier heights (*solid data points*) we find coherent spin squeezing which requires many-body entanglement. The *curved lines* are the boundaries for the indicated amount of coherent spin squeezing. Open symbols represent measurements taken for a high barrier. Number squeezing is slightly larger than for intermediate barriers, but phase coherence is decreased. The measurements lie in the number squeezed region where entanglement is only required for a symmetric quantum state. The *inset* shows our measurements in comparison to the lowest achievable number fluctuations at a given coherence revealing the large effect of increased entropy due to finite temperature. This figure is originally published in [45]

The inset in Fig. 3.19 clarifies the effect of finite temperature on the spin squeezing in our experiment. In contrast to the main figure the vertical axes is rescaled to show the best achievable number squeezing for a given phase coherence (gray line). Ideal measurements on the ground state of a Josephson junction would yield results close to this line. The measured data points are approximately 25 dB above the best achievable coherent spin squeezing showing large room for future improvement if the entropy of the system can be better controlled.

3.4.4 Particle Loss and Number Squeezing

The results on squeezing presented above rely on an accurate calibration of the atom number detection (see also Appendix A). An independent test of the calibration

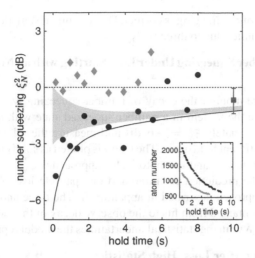

Fig. 3.20 Number squeezing and particle loss. We measure the evolution of number squeezing with particle loss in order to test for our absorption imaging calibration. Data shown as gray diamonds correspond to measurements where we condensed directly into a high barrier trap setup with negligible coupling between the wells. No number squeezing is observed in this case. *Black circles* show measurements where we start from a squeezed state and hold the atoms in the trap for different times. One and three body losses cause a decay of number squeezing. The *black solid line* is the theoretical prediction where the loss coefficients are extracted from the observed total atom number decay shown in the *inset*. The *gray square* data point summarizes the fluctuation measured after 10 s starting with a slightly squeezed state. The particle loss reduces the uncertainty in the number squeezing (*gray shaded area*) which allows for a good quantitative comparison to the theory. In order to obtain a strong test of our calibration we repeat the experiment 1,000 times resulting in small statistical errors. The given error bars represent two statistical standard deviations and show good agreement with the theory. This figure is originally published in [45]

can be done by monitoring the evolution of number squeezing when the system is subject to particle loss. Appendix B details the calculation of the evolution of number squeezing when one and three body loss is present (two body loss is negligible for ^{87}Rubidium in the $|F, m_F\rangle = |2, \pm 2\rangle$ states [49, 50]). We perform three different measurements:

Evolution of Number Squeezing Under Loss Starting with an Uncorrelated State

We prepare a Bose–Einstein condensate in a very high lattice situation ($V_0 = 2\pi \cdot 2,700$ Hz) such that the tunneling time $\tau_{pl} = 2\pi/\omega_{pl}$ between adjacent sites is in the order of a few tens of seconds and therefore longer than the experimental timescale. The condensates on the different lattice sites are independent and we expect poissonian number fluctuations between them. Only small dynamics in the number squeezing versus particle loss is expected due to correlations stemming from three particle loss. The gray diamonds in Fig. 3.20 show the measured number squeezing versus the evolution time. Although we loose approximately two-thirds of the atoms (see inset), the measured data points scatter around $\xi_N^2 = 0$ dB showing that the calibration of our imaging is correct. Due to limited statistics we are not able to observe the dynamics due to three body loss.

Evolution of Number Squeezing Under Loss Starting with an Number Squeezed State

As a further test we monitor the decay of number squeezing when particle loss is present. We prepare the system in a number squeezed state with an initial number squeezing of approximately $\xi_N^2 = -6$ dB and measure the relative atom number fluctuations after different hold times. The result is plotted as solid circles in Fig. 3.20. Number squeezing decays and asymptotically approaches $\xi_N^2 = 0$ dB. The solid black line shows the prediction for three and one particle loss obtained from the Master equation approach described in appendix B. The three and one body decay coefficients where obtained from fits to the observe decay of the mean atom number shown in the inset. Within the statistical uncertainties the model reproduces our data.

Number Squeezing After Loss: High Statistics

The gray square data point in Fig. 3.20 is the strongest test of our imaging calibration. Here we start with a slightly squeezed state by condensing into a $V_0 = 2\pi \cdot 430$ Hz lattice which is the usual starting point for most of the experiments presented above. After a fast barrier ramp up in 20 ms to $V_0 = 2\pi \cdot 2,700$ Hz we expect an initial number squeezing $-3\,\mathrm{dB} < \xi_N^2 < 0\,\mathrm{dB}$ as detected in the measurement shown in Fig. 3.17. We measure the relative atom number fluctuations after 10 s evolution time during which two-thirds of the atoms are lost and we find $\xi_N^2 = -0.7^{+0.7}_{-0.7}$ dB, where the indicated errors are 95% statistical confidence bounds. This has to be compared to the expected fluctuations of $-1.2\,\mathrm{dB} < \xi_N^2 < -1\,\mathrm{dB}$ predicted from the measured loss rates. 1,000 experimental realizations contribute to the measurement which allows for a quantitative comparison with the theory. Within the remaining statistical uncertainties we find good agreement between theory and experiment.

References

1. Bose S (1924) Plancks Gesetz und Lichtquantenhypothese. Z Phys 26:178
2. Einstein A (1924) Quantentheorie des einatomigen idealen Gases. Sitzungsber Kgl Preuss Akad Wiss 261
3. Einstein A (1925) Quantentheorie des einatomigen idealen Gases. Sitzungsber Kgl Preuss Akad Wiss 3
4. Anderson MH, Ensher JR, Matthews MR, Wieman CE, Cornell EA (1995) Observation of Bose–Einstein condensation in a dilute atomic vapor. Science 269:198–201
5. Davis KB et al (1995) Bose–Einstein condensation in a gas of sodium atoms. Phys Rev Lett 75:3969–3973
6. Bradley CC, Sackett CA, Hulet R G (1997) Bose–Einstein condensation of lithium: observation of limited condensate number. Phys Rev Lett 78:985–989
7. Ketterle W (2002) Nobel lecture: when atoms behave as waves: Bose–Einstein condensation and the atom laser. Rev Mod Phys 74:1131–1151

8. Cornell EA, Wieman CE (2002) Nobel lecture: Bose–Einstein condensation in a dilute gas the first 70 years and some recent experiments. Rev Mod Phys 74:875–893
9. Pethik H (2002) Bose–Einstein condensation in dilute gases. Cambridge University Press, Cambridge
10. Pitaevski L, Stringari S (2003) Bose–Einstein condensation. Oxford University Press, Oxford
11. Ketterle W, Durfee D, Stamper-Kurn D (1999) Making, probing and understanding Bose–Einstein condensates. arXiv:cond-mat/9904034v2
12. Feynman R (1982) Simulating physics with computers. Int J Theor Phys 21:467–488
13. Feynman R (1986) Quantum mechanical computers. Found Phys 16:507–531
14. Bloch I, Dalibard J, Zwerger W (2008) Many-body physics with ultracold gases. Rev Mod Phys 80:885
15. Bouyer P, Kasevich MA (1997) Heisenberg-limited spectroscopy with degenerate Bose–Einstein gases. Phys Rev A 56:R1083–R1086
16. Dowling JP (1998) Correlated input-port matter-wave interferometer: quantum-noise limits to the atom-laser gyroscope. Phys Rev A 57:4736–4746
17. Sørensen AS, Mølmer K (2001) Entanglement and extreme spin squeezing. Phys Rev Lett 86:4431–4434
18. Dunningham JA, Burnett K, Barnett SM (2002) Interferometry below the standard quantum limit with Bose–Einstein condensates. Phys Rev Lett 89:150401
19. Pezzé L, Smerzi A (2009) Entanglement, nonlinear dynamics, and the Heisenberg limit. Phys Rev Lett 102:100401
20. Giovannetti V, Lloyd S Maccone L (2004) Quantum-enhanced measurements: beating the standard quantum limit. Science 306:1330–1336
21. Giovannetti V, Lloyd S, Maccone L (2006) Quantum metrology. Phys Rev Lett 96:010401
22. Albiez M et al (2005) Direct observation of tunneling and nonlinear self-trapping in a single bosonic Josephson junction. Phys Rev Lett 95:10402
23. Weller A (2008) Dynamics and interaction of dark solitons in Bose–Einstein condensates. Ph.D. thesis, University of Heidelberg
24. Albiez M (2005) Observation of nonlinear tunneling of a Bose–Einstein condensate in a single Josephson junction. Ph.D. thesis, University of Heidelberg
25. Gati R (2007) Bose–Einstein condensates in a single double well potential. Ph.D. thesis, University of Heidelberg
26. Grimm R, Weidemuller M, Ovchinnikov YB (2000) Optical dipole traps for neutral atoms. Adv At Mol Opt Phy 42:95
27. Gati R, Albiez M, Fölling J, Hemmerling B, Oberthaler M (2006) Realization of a single Josephson junction for Bose–Einstein condensates. Appl Phys B 82:207–210
28. Zapata I, Sols F, Leggett AJ (1998) Josephson effect between trapped Bose–Einstein condensates. Phys Rev A 57:R28–R31
29. Milburn G, Corney J, Wright E, Walls D (1997) Quantum dynamics of an atomic Bose–Einstein condensate in a double-well potential. Phys Rev A 55:4318–4324
30. Ananikian D, Bergeman T (2006) Gross–Pitaevskii equation for Bose particles in a double-well potential: two-mode models and beyond. Phys Rev A 73:13604
31. Cohen-Tannoudji C, Dupont-Roc J, Grynberg G (1992) Atom-photon interactions: basic processes and applications. Wiley, New York
32. Leggett A (2001) Bose–Einstein condensation in the alkali gases: some fundamental concepts. Rev Mod Phys 73:307–356
33. van Oosten D, van der Straten P, Stoof H (2003) Mott insulators in an optical lattice with high filling factors. Phys Rev A 67:33606
34. Spekkens RW, Sipe JE (1999) Spatial fragmentation of a Bose–Einstein condensate in a double-well potential. Phys Rev A 59:3868–3877
35. Raghavan S, Smerzi A, Fantoni S, Shenoy SR (1999) Coherent oscillations between two weakly coupled Bose–Einstein condensates: Josephson effects, π oscillations, and macroscopic quantum self-trapping. Phys Rev A 59:620–633

36. Cohen-Tannoudji C, Diu B, Laloe F (2005) Quantum mechanics. Wiley-VCH, New York
37. Pezzé L, Smerzi A, Berman GP, Bishop AR, Collins LA (2006) Nonlinear beam splitter in Bose–Einstein-condensate interferometers. Phys Rev A 74:033610
38. Bodet C, Estève J, Oberthaler MK, Gasenzer T (2010) Two-mode Bose gas: beyond classical squeezing. Phys Rev 81:063605
39. Pezzé L, Collins LA, Smerzi A, Berman GP, Bishop AR (2005) Sub-shotnoise phase sensitivity with a Bose–Einstein condensate Mach–Zehnder interferometer. Phys Rev A 72:043612
40. Gati R, Hemmerling B, Fölling J, Albiez M, Oberthaler M (2006) Noise thermometry with two weakly coupled Bose–Einstein condensates. Phys Rev Lett 96:130404
41. Pitaevskii L, Stringari S (2001) Thermal vs quantum decoherence in double well trapped Bose–Einstein condensates. Phys Rev Lett 87:180402
42. Benson AK (1975) A procedure for obtaining quantum mechanical transformation of diagonalization from the classical. Int J Theor Phys 12:251–260
43. Gross C, Estève J, Oberthaler MK, Martin AD, Ruostekoski J (2011) Local and spatially extended sub–Poisson atom–number fluctuations in optical lattices. Phys Rev A 84:011609
44. Grubbs FE (1969) Procedures for detecting outlying observations in samples. Technometrics 11:1–21
45. Estève J, Gross C, Weller A, Giovanazzi S, Oberthaler MK (2008) Squeezing and entanglement in a Bose–Einstein condensate. Nature 455:1216–1219
46. Javanainen J, Ivanov M (1999) Splitting a trap containing a Bose–Einstein condensate: atom number fluctuations. Phys Rev A 60:2351–2359
47. Grond J, Schmiedmayer J, Hohenester U (2009) Optimizing number squeezing when splitting a mesoscopic condensate. Phys Rev A 79:021603
48. Grond J, von Winckel G, Schmiedmayer J, Hohenester U (2009) Optimal control of number squeezing in trapped Bose–Einstein condensates. Phys Rev A 80:053625
49. Burt E et al (1997) Coherence, correlations, and collisions: what one learns about Bose–Einstein condensates from their decay. Phys Rev Lett 79:337–340
50. Söding J et al (1999) Three-body decay of a rubidium Bose–Einstein condensate. Appl Phys B 69:257–261

Chapter 4
Non-linear Interferometry Beyond the Standard Quantum Limit

In the experiments described in the previous chapter we detected coherent spin squeezed atomic quantum states. However the implementation of a full atom interferometer where the two modes are defined by two mean field wavefunctions in a double well potential is difficult. One of the problems is the limited range in which the system parameter Λ can be tuned—especially the Rabi regime is not accessible for our setup [1]. Therefore the realization of a beamsplitter, i.e., a $\pi/2$ pulse between the two external modes, remains an open challenge [2]. The second issue is due to the tuning of the Hamiltonian parameters by changing the external trapping potential. Standard coupling pulses in an atom interferometer cause unitary rotations which requires diabatic changes of the hamiltonian parameters. This involves a fast change of the external trapping potential which is—without exiting the system—only possible if the timescale corresponding to the local intra-well trapping frequencies is much faster than the inverse plasma frequency. This requirement is not fulfilled for our setup and fast changes of the barrier height initiates the breathing motion or even dipole motion of the individual condensates.[1]

In the experiments described in this chapter we overcome these problems and we present an interferometric measurement directly demonstrating phase precision beyond the standard quantum limit. The prerequisites necessary to implement the novel non-linear interferometer are described first, and the main result can be found in Sect. 4.7 at the end of the thesis.

An Interacting Two-Mode System Defined by Atomic Hyperfine States

We extend our experimental setup such that two internal hyperfine states of the ^{87} Rubidium atoms are used as the two modes to overcome the limitations mentioned above. The coupling between the two states can be tuned in much cleaner way since it does not require any change of the external trapping potential but electromagnetic

---

[1] We use the excitation of these modes when changing the barrier height V_0 abruptly for a calibration of V_0 [3].

C. Groß, *Spin Squeezing and Non-linear Atom Interferometry with Bose–Einstein Condensates*, Springer Theses, DOI: 10.1007/978-3-642-25637-0_4, © Springer-Verlag Berlin Heidelberg 2012

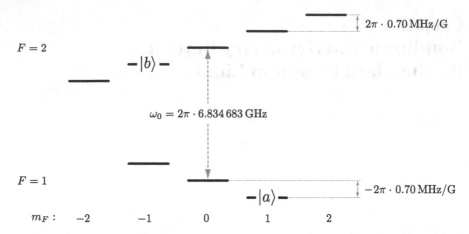

Fig. 4.1 Hyperfine structure of the $5^2 S_{1/2}$ electronic ground state of ^{87} Rubidium. This schematic figure shows the two ground state hyperfine manifolds of ^{87} Rubidium including their Zeeman splitting. The level splittings are not to scale. We work at rather low magnetic fields such that the Zeeman splitting between neighboring states is in the order of $2\pi \cdot 6$ MHz. The two states labeled $|a\rangle$ and $|b\rangle$ form an effective two-level system with almost common mode first order Zeeman shift

radiation is employed. Today's time standard is based on shot noise limited Ramsey interferometry [4] implemented on a similar atomic system—two hyperfine states of Cesium atoms. The microwave technology required for precise coupling pulses is therefore readily available and the experimental techniques necessary to realize standard linear interferometry are well known [5].

However, our goal is to implement high precision non-linear atom interferometry utilizing the interactions between the particles. Therefore the experimental system has to fulfill some specific requirements, which are low sensitivity to magnetic field noise and —most important—finite interaction strength among the atoms. A Bose–Einstein condensate of ^{87} Rubidium has been considered as a promising candidate to create coherent spin squeezing based on two hyperfine states [6]. Figure 4.1 shows the hyperfine structure of ^{87} Rubidium in the electronic ground state. The $|F, m_F\rangle = |1, 1\rangle$ and $|2, -1\rangle$ states in the lower and upper hyperfine manifold are suitable states for this experiment. They fulfill the two major requirements – the tunability of interspecies interactions [7–9] and their magnetic field dependent differential energy shift is small.

4.1 Squeezing: Internal Versus External Degrees of Freedom

In this section we work out the main differences between the external double-well system and two-mode system based on two hyperfine states with respect to the generation of spin squeezed states. It is clarified that the two systems can be described by the same Hamiltonian but also that the experimental limitations are very different.

A fast, diabatic squeezing protocol is shown to be suitable for the realization of coherent spin squeezing based on hyperfine states of ^{87} Rubidium.

4.1.1 The Spin Model

In Sect. 3.1.3 different theoretical descriptions of the external system assuming various approximations have been discussed. On this level of precision, where higher order corrections to the Hamiltonian are neglected [1] the internal system is described by the same Hamiltonians [10]. The main assumption in the internal case is that the spatial wavefunctions of the two hyperfine states are identical [11–13], an issue discussed in detail below. Distinct to the external double well case the Hamiltonian parameters can be switched very fast and also the Rabi regime is easily reached. Therefore arbitrary rotations on the Bloch sphere are possible and detection of any spin direction can be done by a proper unitary rotation of the state before readout. The spin component to be measured is rotated to the J_z direction which is detected by the occupation difference n between the two states [14]. This is a major difference to the experiments discussed in the previous chapter, where phase readout was done by observation of a spatial interference pattern.

The spin Hamiltonian provides the most intuitive description in the internal case

$$H = -\Delta\omega_0 \hat{J}_z + \chi \hat{J}_z^2 - \Omega \hat{J}_x \tag{4.1}$$

where we use the parameters χ for the nonlinearity and the Rabi frequency Ω instead of the Josephson parameters E_C and E_J. The motivation for this nomenclature becomes clear throughout this chapter but, in brief, we use a diabatic technique switching the Hamiltonian parameters such that either the Rabi frequency dominates or the coupling is switched off $\Omega = 0$ such that the evolution is purely due to the non-linear term.

The Coupling Ω

The Hamiltonian parameters are revealed differently than in the external case. The coupling between the two modes $|a\rangle := |1, 1\rangle$ and $|b\rangle := |2, -1\rangle$ is purely given by single particle physics and its strength–the Rabi frequency Ω—is controlled by the intensity of the electromagnetic radiation. The energy of the two states differs by approximately $\omega_0 = 2\pi \cdot 6.8$ GHz and their Zeeman quantum number m_F is distinct by two. We use a similar scheme as described in references [8, 15, 16] to couple the two states by a two-photon transition as shown in Fig. 4.2. We choose a single-photon detuning $\delta = -2\pi \cdot 200$ kHz to the $|2, 0\rangle$ intermediate state allowing for maximal two-photon Rabi frequencies $\Omega = 2\pi \cdot \mathcal{O}(1 \text{ kHz})$. Our experiments require an offset magnetic field of approximately $B_0 = 9.1$ G resulting in a Zeeman shift of ca. $2\pi \cdot 6.3$ MHz between two neighboring Zeeman sub-states in the same hyperfine manifold. Therefore the two-photon pulses comprise of two frequencies,

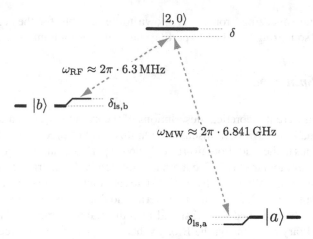

Fig. 4.2 Coupling of the internal two-mode system .This schematic shows the relevant three level scheme necessary to implement the Rabi coupling Ω between the states $|a\rangle$ and $|b\rangle$. Two electromagnetic radiation fields ω_{MW} and ω_{RF} are used to couple the two states with a single-photon detuning δ to the $|2, 0\rangle$ intermediate state. The individually off resonant photon fields cause a light shift of the Zeeman levels here represented as the effective shifts $\delta_{ls,a}$ and $\delta_{ls,b}$. These shifts are linearly dependent on the intensity of the two electromagnetic fields, especially they drop to zero when the fields are switched off resulting in a detuning of the two-photon transition during the free evolution time

one in the microwave regime around $\omega_{MW} = 2\pi \cdot 6.841$ GHz and one in the radio-frequency regime around $\omega_{RF} = 2\pi \cdot 6.3$ MHz. We stress a big difference to the external case—the coupling can be switched from maximum to exactly zero and vice versa faster than any other timescale in the experiment.

Detuning: Rotation Around the J_z Axis

The first term in Equation (4.1) proportional to $\hat{J}z$ describes a rotation of the state around the J_z axis of the Bloch sphere. A priori the angular frequency is given by the energy difference ω_0 between the two states. However the position of the state on the Bloch sphere is measured in a rotating frame whose angular frequency is given by the joint frequency of the two-photon coupling. For resonant pulses this matches exactly the energy splitting between the two modes such that there is no relative rotation in the resonant case [17]. However, for a two-photon coupling scheme the two states experience a differential light shift due to the single-photon detuning δ (see Fig. 4.2). This light shift involves contributions from several of the hyperfine states since we are not dealing with an isolated three level problem. It can be measured accurately by Ramsey spectroscopy where alternately one of the electromagnetic radiation fields is present during the evolution. For the typical microwave and radio-frequency power used in our experiments it is in the order of $\delta_{ls,a} + \delta_{ls,b} = -2\pi \cdot 150$ Hz. This effect leads to a different resonance frequency ω_0 whether the coupling is on or off. Since accurate unitary rotations require resonant coupling, finite rotation dynamics $\Delta\omega_0$ around the J_z axis during free evolution is not avoidable.[2]

Two other effects cause a differential energy shift of the two modes, of which one—the second order Zeeman shift—is a single particle effect and the second— a mean field interaction induced chemical potential difference between the two condensates—is a many particle effect. Mainly the first effect causes experimental difficulties when preparing spin squeezed states since the magnetic field is not controllable to an arbitrary precision and its fluctuations result in excess phase noise. For a more detailed discussion see Sect. 4.5 The analogy to the detuning $\Delta\omega_0$ in the external double well experiments is described in In Sect. 3.3 where relative position fluctuations of the different trapping beams were found to be most critical.

Miscibility and the Nonlinearity χ

Calculation of the nonlinearity χ requires the knowledge of the mean field wavefunctions $\phi_{a,b}$ of the two modes $|a\rangle$ and $|b\rangle$.[3] From the two-mode ansatz (see In Sect. 3.1.3) used to derive Equation (4.1) the nonlinearity follows [6, 10, 11, 13]

$$\chi = \frac{g_{aa}}{2} \int d\mathbf{r} \, |\phi_a|^4 + \frac{g_{bb}}{2} \int d\mathbf{r} \, |\phi_b|^4 - g_{ab} \int d\mathbf{r} \, |\phi_a|^2 \, |\phi_b|^2$$

$$\approx \frac{1}{2}(g_{aa} + g_{bb} - 2g_{ab}) \int d\mathbf{r} \, |\phi_a|^4 \tag{4.2}$$

with coupling constants $g_{ij} = 4\pi\hbar^2 a_{ij}/m$ and s-wave scattering lengths a_{ij} between states i and j. The mean field wavefunctions of each mode are normalized to unity $\int d\mathbf{r} \, |\phi_i|^2 = 1$. The last approximation assumes equal spatial mean field wavefunctions for both modes. The same expression can be derived in the external double well case [18, 19] for the calculation of the charging energy $E_C/2$, however the big difference is that the overlap of the two mean field wavefunctions is small in this case such that the third term in the first line of Equation (4.2) almost vanishes. In the external case the last approximation therefore certainly does not hold.

For [87] Rubidium and the chosen hyperfine states the background s-wave scattering lengths a_{ij} are almost equal [7, 8, 12]

$$a_{aa} = 100.44 a_B$$
$$a_{bb} = 95.47 a_B \tag{4.3}$$
$$a_{ab} = 97.7 a_B$$

where a_B is the Bohr radius. Therefore the effective scattering length is close to zero $a_{aa} + a_{bb} - 2a_{ab} = 0.5 a_B$ resulting in a negligible nonlinearity $\chi = 2\pi \cdot \mathcal{O}(10^{-3}\,\text{Hz})$.

---

[2] Technologically a coherent change of the frequency of one of the two electromagnetic fields is not possible in our experiment.

[3] We use the labeling $|a\rangle$ and $|b\rangle$ for both the single particle states and the mean field modes which is not rigorously correct. However since we neglect external dynamics and assume perfect wavefunction overlap this labeling is justified.

There are two possibilities to increase the nonlinearity χ, one involves independent control of the external potentials seen by the two states such that the wavefunction overlap can be tuned [12, 20], the second is to make use of a magnetic Feshbach resonance which provides a handle to control the interspecies s-wave scattering length a_{ab} [21]. The latter method is suitable for our experimental setup since we work with optical dipole traps where Feshbach resonances can be readily used, but state selective potentials are hard to implement. The ratios of the background scattering lengths given above cause another difficulty. The two condensates tend to de-mix [22] since the miscibility condition $a_{ab}^2 < a_{aa}a_{bb}$ [23] is not fulfilled. Depending on the external trapping of the Bose–Einstein condensates this leads to a breakdown of the two-mode approximation and the Hamiltonian (4.1) is no longer appropriate to describe the system. However for the chosen states the system is very close to the miscible regime and a small decrease of 0.3% of the interspecies scattering length a_{ab} would ensure miscibility.

4.1.2 Interaction Tuning via A Magnetic Feshbach Resonance

A suitable interspecies Feshbach resonance between the $|1, 1\rangle$ and $|2, -1\rangle$ states around $B_0 = 9.10$ G has been reported [7–9] and tunability of the interstate scattering length a_{ab} in the order of 10% has been shown [8]. Figure 4.3 shows the theoretical prediction of the dependence of the scattering length from the magnetic field

$$a_{ab}(B) = a_{ab}^{\text{bg}}\left(1 - \frac{\delta B}{(B - B_0 - i\gamma_B/2)}\right) \qquad (4.4)$$

with a resonance width of $\delta B = 2.0$ mG and a decay width of $\gamma_B = 4.7$ mG accounting for enhanced inelastic spin relaxation and three body loss [7] ($a_{ab}^{\text{bg}} = 97.7a_B$ is the background value). Thus—distinct to the experiments on external degrees of freedom where no suitable Feshbach resonance exists in the low magnetic field regime—the nonlinearity χ between the two hyperfine states can be rather easily tuned. The maximum achievable nonlinearity is limited by the enhanced inelastic losses when working close to the Feshbach resonance (Fig. 4.3b) [21]. Furthermore, in a spin squeezing experiment these losses limit the maximum achievable correlations between the spin directions [24]. Therefore a balance between elastic and inelastic enhancement of the scattering properties is important and the optimal magnetic field for our experiments is chosen taking this problem into account (see Sect. 4.1.3).

Figure 4.3a reveals that the elastic interspecies scattering length is lowered for $B > B_0$ causing the system to enter the miscible regime. Nevertheless the single spatial mode approximation—both hyperfine states share the same spatial wavefunction—does not necessarily hold in a dynamic experiment since a sudden change of the hyperfine state, as done by a $\pi/2$ pulse, initiates external dynamics in the system due to the different mean field potentials seen by the two states ($a_{aa} \neq a_{bb}$). However the effect of these dynamics has shown to be negligible for our optical trap configuration

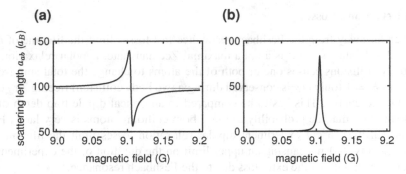

Fig. 4.3 Theoretical prediction of the interspecies Feshbach resonance. Panel **a** shows the dependence of the elastic part of the scattering length—the real part of Equation (4.4)—around a magnetic field of $B_0 = 9.10$ G. The interspecies scattering length is decreased for $B > B_0$ and the system enters the miscible regime. The boundary for a_{ab} between the miscible and non-miscible regime is shown by the *gray line* revealing miscibility in a large range above the resonance. In **b** the Lorenzian shaped inelastic part of the scattering length—the imaginary part of Equation (4)—is plotted

(see Sect. 4.5) [12] and in principle it could be even further suppressed by working in the *breath together* regime as proposed in [11, 12].

Due to the small elastic width of the resonance absolute stability of the magnetic field in the order of a few milligauss is required. We can achieve this stability on intermediate timescales of a few hours after which slow thermally induced drifts cause a change of the magnetic field. We measure drifts of approximately 5 mG on the timescale of one day. Technical constraints of our active magnetic field stabilization forbid fast changes in the magnetic field during the experimental sequence (see Appendix C). Therefore the two-photon coupling pulses have to be done close to the Feshbach resonance and effects of radio-frequency dressing of molecular states can become important depending on the frequency of the radio-frequency coupling field [7]. However, for the chosen detuning $\delta = -2\pi \cdot 200$ kHz we see none of these effects.

4.1.3 Experimental Characterization of the Feshbach Resonance

Magnetic Feshbach resonances allow for convenient tuning of the interaction strength in experiments with ultracold atoms. Not only elastic collision properties are altered but in most cases also the inelastic collision rate increases [21]. A loss rate measurement is performed in order to determine the enhanced inelastic collision rate close to the resonance while mean field spin dynamics can be employed to measure the effect of elastic collisions [25].

Spin Relaxation Losses

Even far away from the Feshbach resonance fast losses limit the lifetime of the $|b\rangle = |2, -1\rangle$ state since it is a not a maximal Zeeman state. Dipolar relaxation in two body collisions causes one or both of the atoms to change the total spin from $F = 2$ to $F = 1$ (only m_F is conserved during a two body collision) freeing $2\pi \cdot 6.8$ GHz kinetic energy. This has to be compared to an optical dipole trap depth of a few kHz, such that the probability to loose both colliding atoms is very large. For typical densities in our experiment dipolar relaxation losses limit the lifetime to approximately 250 ms causing an upper limit on the duration of the experimental sequence even without the extra loss due to the Feshbach resonance.[4]

The interspecies Feshbach resonance between the two hyperfine states has a rather large inelastic width caused by enhanced two body collisions—dipolar relaxation losses—and three body collisions involving molecule formation [21]. We perform a measurement of the magnetic field dependent loss rate around the interspecies resonance centered at $B_0 = 9.10$ G. For the estimation of the loss rate we fit the observed atom number decay $N_i(t)$ with an exponential model $N_i(t) = N_i(0)e^{-t/\tau_{\text{loss}}}$ where $i = a$ or b (Fig. 4.4(a)). This is an approximation since the loss rate τ_{loss}^{-1} due to two and three body loss is density dependent and the decay is not exactly exponential. Figure 4.4b summarizes the result of this measurement. The Lorenzian fit reveals a inelastic width of 11.7 mG.[5]

At $B = 9.13$ G the loss rate is $\tau_{\text{loss}}^{-1} \approx 10$ Hz resulting in a loss of 10%–15% of the atoms after 20 ms, the typical timescale for a diabatic squeezing experiment (see In Sect. 4.2)). The loss limits the number squeezing to approximately $\xi_N^2 = -10$ dB, which is close to the theoretical optimum achievable for our experimental parameters, i.e. total atom number and external trap configuration [12]. Therefore we choose $B = 9.13$ G as our working point.[6]

The rather short lifetime of the condensate in a superposition of the two internal states due to the losses described above is another difference to the external squeezing experiments in a double well potential. It is the main limitation to obtain larger spin squeezing in the internal system and restricts the available methods to obtain a reasonable amount of spin squeezing (see Sect. 4.2)

Tuning of the Nonlinearity χ

Despite of the strong inelastic collisions it is still possible to tune the nonlinearity χ in a useful way. Mean field spin dynamics are used to extract the effective nonlinearity which is measured from the frequency difference between small amplitude

---

[4] Dipolar relaxation is a two body process meaning its rate $L_2 \propto K_2 N$ is proportional to the number of atoms in the trap. The loss coefficient for the $|2, -1\rangle$ state was measured to $K_2 = 8.8 \times 10^{-14}$ cm^3/s [26]

[5] At the time this measurement was done the active magnetic field compensation was not yet installed. Therefore we measure a larger width as theoretically expected. The inelastic width extracted from the measurement shown in Fig. 4.5 agrees well with the theoretical prediction.

[6] Experimentally we also found the best number squeezing at this magnetic field.

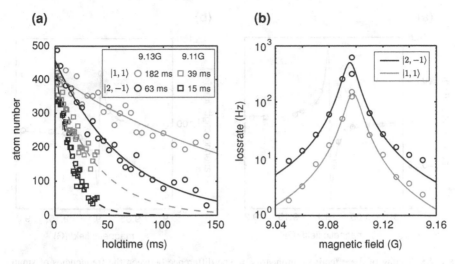

Fig. 4.4 Loss rate measurement in the vicinity of the Feshbach resonance. **a,** Exemplary loss curves for both hyperfine states and two magnetic fields are shown. The *dashed and solid lines* are exponential fits to the data from which the loss rates are extracted. The respective life times are given in the legend of the figure. Panel **b** shows the loss rate in the vicinity of the Feshbach resonance including a Lorenzian fit to the data for both hyperfine states. The greater loss rate of the $|2, -1\rangle$ state with respect to the $|1, 1\rangle$ state is explained by additional dipolar relaxation losses in the $|2, -1\rangle$ state. The slight offset between the center of the two fits is not significant—within the fit uncertainties the two centers match

oscillations around zero relative phase and π relative phase [25]. Experimentally a coherent spin state $|\theta = \pi/2, \varphi_0\rangle$ is prepared on the equator but with small offset relative phase such that $\varphi_0 = 0 + \epsilon$ or $\varphi_0 = \pi + \epsilon$ where $\epsilon \approx 0.1 \cdot \pi$. This initial quantum state evolves under the Josephson Hamiltonian (4.1) and at time t the quantum state is:

$$|\theta, \varphi\rangle(t) = e^{-it(\chi \hat{J}_z^2 - \Omega \hat{J}_x)}|\theta = \pi/2, \varphi_0\rangle \qquad (4.5)$$

Sinusoidal oscillations in the population imbalance $\langle \hat{J}_z(t)\rangle$ versus time are observed and their respective frequency $\omega_{pl,\pi}$ is extracted. In mean field approximation analytic expressions for the expected oscillation frequencies have been derived [25]:

$$\omega_{pl,\pi} = \Omega\sqrt{1 \pm \Lambda} \qquad (4.6)$$

The difference of these two frequencies reveals the effective nonlinearity χ since the Rabi frequency Ω and the atom number N are known:

$$\chi = \frac{\Omega \Lambda}{N} = \frac{\left(\omega_{pl}^2 - \omega_\pi^2\right)}{2N\Omega} \qquad (4.7)$$

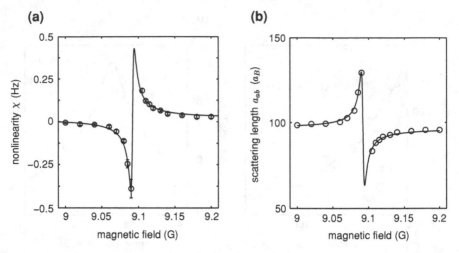

Fig. 4.5 Tuning of the effective nonlinearity. **a** The difference between the frequencies of small amplitude oscillations around zero and χ relative phase reveal the nonlinearity χ. The *black solid line* is a fit assuming the real part of Equation (4.4) as the functional dependence between elastic collision enhancement and magnetic field. Panel **b** shows the deduced interspecies scattering length a_{ab} in the vicinity of the Feshbach resonance. From the fit we obtain an inelastic width of $\gamma_B = 4.6 \pm 0.7$ mG in accordance with the theoretical prediction. Compared to the direct measurement of the loss rate enhancement presented in Fig. 4.4 the active magnetic field stabilization was installed in this measurement which explains the difference in the observed widths. The elastic width extracted from the fit is $\delta B = 1.6 \pm 0.2$ mG. We find the center of the resonance at $B_0 = 9.092$ G, but since we did not focus on a precise calibration of the absolute magnetic field we estimate a systematic error of approximately 10 mG on this value. The background scattering length a_{ab}^{bg} was chosen as a free parameter and the fit reveals $a_{ab}^{bg} = 96.5 \pm 0.7 a_B$

Figure 4.5a shows the deduced nonlinearity around the Feshbach resonance. The evolution was measured for a Rabi frequency of $\Omega = 2\pi \cdot 200$ Hz[7] and for the same atom numbers as used in the following experiments. We find $\chi = 2\pi \cdot 0.063$ Hz at $B = 9.13$ G in accordance with simulations using the Gross-Pitaevskii equation.

In Fig. 4.5b we deduce the interspecies scattering length a_{ab} from the measured nonlinearity using equation (4.2). We calculate the integral $\int d\mathbf{r} |\phi_a|^4$ for the wavefunction ϕ_a obtained from the numerical solution of the Gross-Pitaevskii equation. Due to this procedure we estimate a possible systematic error of 10% in the data shown in Fig. 4.5b

4.1.4 What About Temperature?

Finite temperature and entropy is the limiting factor for the squeezing experiments based on two mean field modes of a Bose–Einstein condensate in an external dou-

---

[7] A lower Rabi frequency than for the usual coupling is chosen here in order to work in a higher Λ situation.

ble well potential. In Sect. 3.2 an argument was given that thermal excitation of the dipole mode in trap-splitting direction translates to increased entropy in the Josephson many-body modes which limits coherent spin squeezing. The situation is very different when two internal states are used. As described later, the experiments start with a system in a maximal Dicke state—only state $|a\rangle$ is populated while the second $|b\rangle$ is exactly empty. We ensure this starting condition by pulsing a resonant laser pulse to remove possible population in the upper hyperfine manifold prior to the squeezing experimental sequence. The thermal energy scale is seven orders of magnitudes smaller than the energy difference between the two hyperfine states resulting in negligible thermal excitation initially. After preparing a coherent spin state with finite population in both modes thermal effects might become an issue through a coupling to the mean field dynamics of the condensates. However during the short timescale of our experiment—approximately 20 ms—we find no sign of thermalization and the all experimental results are explained by a zero temperature two-mode model.

The fraction of thermal atoms is also negligible since we work at very low temperatures $T/T_C \approx 1/15$ resulting in a thermal fraction of $1 - N_0/N \approx 10^{-3}$—approximately one atom out of $1,000$ is not in the condensate.

4.2 Fast Diabatic Spin Squeezing by One Axis Twisting Evolution

Adiabatic Protocol

In the external squeezing experiment we used an adiabatic scheme to generate spin squeezing. In principle this is also possible for squeezing based on internal degrees of freedom, but given the combination of rather fast losses and small nonlinearity, it is—even at zero temperature—not the optimal way. This becomes clear when looking at the occupation number fluctuations for the Josephson ground state given in Equation (3.16) which can be used to express the number squeezing ξ_N^2 in dependence of the regime parameter Λ

$$\xi_N^2 = \frac{4\Delta n^2}{N} = \sqrt{\frac{1}{1+\Lambda}} \qquad (4.8)$$

and $\Lambda = N\chi/\Omega$. Typical values for the parameters are $N\chi = 2\pi \cdot 50$ Hz which requires to reduce Ω from approximately $2\pi \cdot 500$ Hz adiabatically to $2\pi \cdot 3$ Hz in order to achieve $\xi_N^2 \approx -6$ dB. In order to check the required ramp time for an adiabatic evolution (assuming a linear ramp) we perform a numerical simulation within the two-mode approximation (Fig. 4.6). The measured lifetime of the atoms is approximately 250 ms and linear ramps of that duration are not yet adiabatic but the number squeezing is only $\xi_N^2 \approx -4$ dB.

No particle losses are taken into account in Fig. 4.6, but assuming—as an estimation—only single particle loss from a squeezed state with $\xi_N^2 \approx -6$ dB, loss

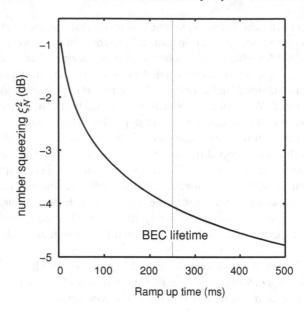

Fig. 4.6 Feasibility of adiabatic number squeezing. In order to estimate the possibility to reach significant spin squeezing close to the ground state of the Josephson Hamiltonian we performed a zero temperature two-mode simulation for our parameters. Starting with a spin state on the equator of the Bloch sphere we reduce the Rabi frequency from $2\pi \cdot 500$ Hz to $2\pi \cdot 3$ Hz by linear ramping with different total ramp time. Within the lifetime of the condensate (marked by the *gray line*), we find that the ramping is not yet adiabatic and only approximately $\xi_N^2 \approx -4$ dB of number squeezing can be reached. Estimation of the effect of particle losses reduces this number even further to approximately $\xi_{N,\text{loss}}^2 \approx -2$ dB (see main text)

of 50% of the atoms degrades the squeezing to $\xi_{N,\text{loss}}^2 \approx -2$ dB. These arguments show that a different scheme is necessary to achieve significant number squeezing for our parameters.

Diabatic Protocol: One Axis Twisting

Particle loss clearly limits the experimental time available to achieve spin squeezing. Employing dynamic strategies spin squeezing can be produced much faster than with adiabatic techniques. One example is the 1993 proposed *one axis twisting* scheme which uses non-linear phase dispersion as the basic mechanism [27] and it has been already considered useful for Bose–Einstein condensates [6, 13]. The scheme is similar to *Kerr effect* based squeezing protocols in quantum optics where a material with a Kerr nonlinearity is used such that the refractive index $n_{\text{light}} \propto n_2 |E|^2$ is proportional to the light intensity $|E|^2$. The light experiences intensity dependent phase modulation within this medium resulting in quadrature squeezing [28].

In the original one axis twisting scheme for atoms, a coherent spin state $|\theta = \pi/2, \varphi\rangle$ evolves for a given time τ under the Hamiltonian:

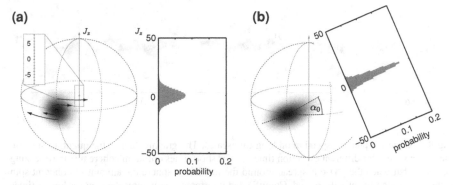

Fig. 4.7 One axis twisting evolution. Panel **a** illustrates the initial coherent spin state on the equator of the Bloch sphere. The *histogram* shows the binomial probability distribution over the Dicke states. The zoom window illustrates the quantization on the J_z axis—the different Dicke states—of which each eigenstate m rotates with a different angular frequency $m \cdot \chi$ around the vertical axis. Part **b** illustrates the quantum state after a short evolution time $\tau < \tau_{deph}$. The isotropic uncertainty developed into an elliptical one with spin squeezing present under an angle α_0. The *histogram* shows the squeezed probability distribution over the eigenstates in a coordinate system rotated by α_0

$$\hat{H} = \chi \hat{J}_z^2 \tag{4.9}$$

Given this Hamiltonian, the time evolution of any quantum state is determined by the unitary operator

$$\hat{U}(t) = e^{-it\chi \hat{J}_z^2} \tag{4.10}$$

which describes a J_z dependent rotation around the J_z axis. In Sect. 2.2.1 coherent spin states were introduced in the first quantization formalism. It was pointed out that a coherent spin state on the equator of the Bloch sphere $|\theta = \pi/2, \varphi\rangle$ can be described as a coherent superposition of several Dicke states where the probability distribution over these basis states is binomial. Within this picture the one axis twisting evolution (Equation (4.10)) of an initial coherent spin state can be nicely visualized. Each Dicke state $|J, m\rangle$ composing the coherent spin state rotates with a different frequency around the J_z axis where the difference in rotation frequency between next neighboring Dicke states is χ (Fig. 4.7a).

For short evolution times $\tau < \tau_{deph}$ this shearing effect results in spin squeezing under an axis rotated by the angle $\alpha_0(\tau)$ with respect to the equator of the Bloch sphere (Fig. 4.7 b). $\tau_{deph} = (\sigma_m \chi)^{-1}$ is the dephasing time, after which the coherence $\langle \cos(\varphi) \rangle \approx 2/N\langle \hat{J}_x \rangle$ [8] has dropped from a value close to unity to $\langle \cos(\varphi) \rangle = e^{-1}$ due to the interaction induced spread of the state around the Bloch sphere [30]. In

---

[8] We assume here without loss of generality $\langle \hat{J}_y \rangle = 0$ such that the twist is symmetric to the J_y axis.

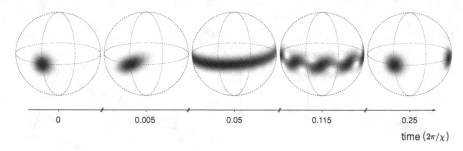

0	0.005	0.05	0.115	0.25

time $(2\pi/\chi)$

Fig. 4.8 Coherent phase dispersion on long timescales. The graphic depicts the quantum state on the *Bloch sphere* for different evolution times τ. For short times optimum coherent spin squeezing develops, but when the phase is spread around the full Bloch sphere the amount of coherent spin squeezing is significantly decreased. Complicated interference structures develop for longer times and the last *Bloch sphere* illustration shows a partial revival of the wavefunction at $\tau = 1/4 \cdot 2\pi/\chi$. For an odd number of atoms the dynamics is periodic with period duration $2\pi/\chi$ while for an even number of particles a phase of π distinguishes the initial quantum state from the quantum state $t = 2\pi/\chi$ and the revival period is twice as long [29]. The calculation was done for 100 atoms

general τ_{deph} is inversely proportional to the extension of the quantum state over the Dicke basis σ_m which, for a coherent spin state on the equator of the Bloch sphere, is $\sigma_m = \sqrt{N}/2$.

At evolution times $\tau \gtrsim \tau_{deph}$ reduced quantum fluctuations under a certain axis are still present but the coherence is very low such that the quantum state is no longer optimally coherently spin squeezed. After even longer times the dynamics show highly nonclassical interference effects [31– 33] finally resulting in a revival of the coherence at $\tau_{rev} = 2\pi \chi^{-1}$ when each neighboring pair of Dicke states is in phase again [32]. Figure 4.8 visualizes the quantum state on the Bloch sphere for different evolution times τ.

The best achievable noise suppression in general interferometry still increases with evolution time even if coherent spin squeezing degrades. It has been recently shown [34] that a new type of Bayesian interferometer readout can be employed to make use of these quantum states (see also Sect. 2.4). Nevertheless this is beyond the scope of this thesis since we focus on standard readout of the population imbalance as used in most of today's Ramsey interferometers.

Timescales to Achieve Squeezing: Adiabatic Versus Diabatic

The maximum squeezing achievable with the diabatic one axis twisting technique is [6, 27]:

$$\xi_s^2 = \frac{3^{2/3}}{2} N^{-2/3} \tag{4.11}$$

This has to be contrasted to the maximum squeezing that can be generated using the adiabatic technique which is given by $\xi_s^2 \approx 2N^{-1}$ at the boundary to the Fock regime where the coherence is still reasonable high [35]. The maximum squeezing is better in the adiabatic case however the adiabaticity criterion requires evolution times $\tau_{\text{adiab}} \approx 1/\chi$. The best squeezing in the one axis twisting protocol is achieved after $\tau_{dia} \approx N^{-2/3} \cdot 1/\chi$, a factor of $N^{2/3}$ faster favoring the diabatic protocol.

The timescales given here are not taking particle loss into account. Losses limit the maximum achievable squeezing and the optimum is reached after a shorter time as compared to the lossless case [12]. For our experimental parameters ($N \approx 500$, $\chi \approx 0.1$ Hz) we expect the best diabatic spin squeezing including particle losses to be in the order of $\xi_s^2 \approx -10$ dB after a non-linear evolution time of $\tau_{dia} = \mathcal{O}$ (10ms).

Analytic Expression for the Variance of the Twisted Quantum State

The initial quantum state—a coherent spin state $|\theta = \pi/2, \varphi\rangle$—features isotropic variance in the directions perpendicular to the mean spin vector. After a certain evolution time τ under the one axis twisting Hamiltonian (4.10) correlations between the two orthogonal directions have been built up and the variance of the spin state is no longer isotropic–the two dimensional variance has an elliptical shape.[9]

Experimentally the fluctuations of the quantum state in J_z direction can be measured and arbitrary unitary rotations of the quantum state are possible. The normalized variance ξ_N^2 in J_z direction after an unitary rotation α around the center of the quantum state (see Fig. 4.9) has been calculated analytically [27]:

$$\xi_N^2(\alpha) = 1 + \frac{N-1}{4}\left[A - \sqrt{A^2 + B^2}\cos(2(\alpha + \delta))\right] \tag{4.12}$$

where the following abbreviations are used:

$$
\begin{aligned}
A &= 1 - \cos(2\chi t)^{N-2} \\
B &= 4\sin(\chi t)\cos(\chi)^{N-2} \\
\delta &= \frac{1}{2}\arctan\left(\frac{B}{A}\right)
\end{aligned}
\tag{4.13}
$$

The required rotation of the quantum state before readout of the population imbalance J_z has been implemented in our lab. Therefore an experimental characterization of the quantum state after one axis twisting evolution is possible and the measurements can be compared to the analytic expression (4.12). The following Sect. 4.3 describes this experiment in detail.

Prior to our experiments this *noise tomography* technique has already been performed experimentally [36, 37]. Coherent spin squeezing and an anisotropic variance distribution have been found for a hot atomic sample in a vapor cell. The experimental

---

[9] We assume small fluctuations $\Delta J_{\perp,\text{max}}^2$ as compared to the total atom number $\Delta J_{\perp,\text{max}}^2 < N^2/4$ such that the Bloch sphere can be locally approximated by a plane.

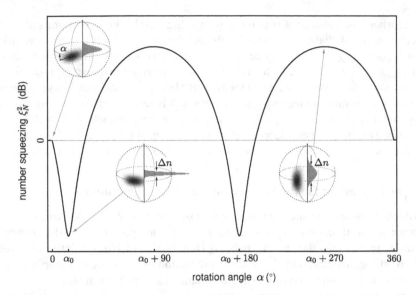

Fig. 4.9 Variance of the J_z spin component versus rotation angle. The figure shows the prediction of Equation (4.12) for the normalized variance ξ_N^2 in J_z spin direction versus rotation angle α around the center of the spin state. The variance is symmetric in intervals $0 \leq \alpha < 180°$ and optimal number squeezing is detected for a rotation angle α_0. At $\alpha_0 + 90°$ the maximally uncertain axis is rotated to the J_z direction. The *Bloch sphere* illustrations show the quantum state and the probability distribution in J_z prior to readout for no rotation, in the number squeezed region and when the anti-squeezed direction is rotated to the vertical axis

protocol used to generate spin squeezing was a quantum non-demolition measurement based method and not the interaction dependent one axis twisting method presented here.

4.3 One Axis Twisting in Action: Experiments

For an internal Josephson contact in a Bose–Einstein condensate *one axis twisting* and *noise tomography* can be experimentally realized by an interferometric sequence similar to the standard Ramsey scheme [38, 39]. In the following we report on these experiments.

Adiabatic Transfer: The Landau-Zener Sweep

Starting with [87] Rubidium atoms in a magneto-optical trap, all our experiments require evaporative cooling in a magnetic time-orbiting-potential trap prior to the transfer of the atoms into the optical dipole traps [40]. Final evaporation is done in the optical traps, defining the temperature and atom number of the Bose–Einstein condensate in a repeatable manner. However, magnetic trapping requires collision-ally stable low field seeking states [41] which are in the case of ^{87} Rubidium the $|F, m_F\rangle = |1, -1\rangle$ and $|2, 2\rangle$ hyperfine states. Different to the external squeez-ing experiment we prepare the Bose–Einstein condensate now in the $|1, -1\rangle$ state. A radio-frequency Landau-Zener sweep within the $F = 1$ hyperfine manifold is used to transfer the atoms with very high efficiency to the state $|a\rangle = |1, 1\rangle$. The radio-frequency is linearly ramped from $2\pi \cdot 1.5$ to $2\pi \cdot 2.5$ MHz in 20 ms realizing the sweep at a moderate magnetic field of $B = 3.2\,\text{G}$.[10]

Interferometric One Axis Twisting Sequence

Here we present the specific implementation of the one axis twisting idea for our experiment. While the magnetic field is actively stabilized at $B = 9.13$ G (see Appendix C) we prepare a coherent spin state by a fast $\Omega t = \pi/2$ pulse. The Rabi frequency is $\Omega \approx 2\pi \cdot 600$ Hz such that the nonlinearity can be neglected during the pulse (Rabi regime with $\Lambda = 0.08$). After this first coupling pulse the phase of the coherent spin state is defined to $\varphi = 0$, meaning the center of mass of the spin state is located on the J_x axis. The state evolves under the Hamiltonian (4.9) for a time τ, symmetrically interrupted by a spin echo pulse after $\tau/2$, until another rotation pulse with appropriately chosen phase is used to rotate the spin state. The pulse phase of the last pulse is adjusted to assure rotation around the center of the spin state and the rotation angle $\alpha = \Omega t_\alpha$ is set by the duration of the pulse t_α. The main external noise source in our experiment are magnetic field fluctuations and we choose the axis of the spin echo pulse such that is perpendicular to the spin polarization direction for lowest noise sensitivity (see Appendix C).[11]

Here it is important to note that there are two possibilities to control the longi-tudinal rotation axis of the coupling pulses: The first one is to set the state rotation angle $\varphi(t)$ via a controlled detuning $\Delta\omega_0$ such that $\varphi(t) = -\Delta\omega_0 t$ at time t. In our setup it is more convenient to choose a second method where we use the fact that the phase φ is defined relative to the phase of the combined microwave and radio-frequency radiation field. Therefore the longitudinal position of the quantum state on the Bloch sphere or equivalently the pulse rotation axis can be chosen by the phase of the subsequent coupling pulses.

In order to perform noise tomography we repeat the experiment 60 times[12] for each angle α and extract the number squeezing ξ_N^2 for each dataset. For an evolution time of $\tau = 18$ ms we find the optimal number squeezing and the results are plotted in Fig. 4.10. For details of the calculation of ξ_N^2 from the raw data see In Sect. 3.4.1. A graphical representation of the tomography sequence can be found in Fig. 4.20

---

[10] The single-photon Rabi frequency for the coupling of the Zeeman sub-states is approximately $2\pi \cdot 10$ kHz.

[11] Experimentally the phase of the coupling pulses at time t can be found by a measurement of the population imbalance versus pulse phase of a final $\pi/2$ pulse. The zero crossings identify the two phases where the rotation axis hits the center of the spin state.

[12] 60 experimental repetitions define one dataset in all measurements done in context with the internal spin system.

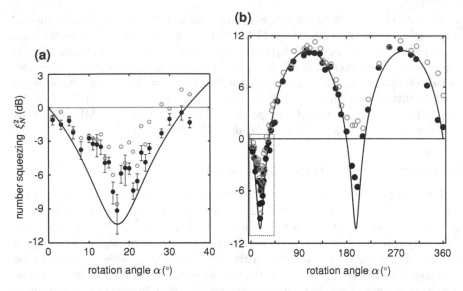

Fig. 4.10 Noise tomography. The figure shows the result of the noise tomography experiment. The *gray open circles* represent the photon shot noise corrected data, while the *black solid* data points are additionally corrected for known technical noise. Panel **a** is a close up of the number squeezed region (the *dashed area* in panel **b**), where for clarity statistical error bars have been added for the *black solid* data only. The *black line* is the theoretical prediction from Equation (4.12) where all parameters have been extracted from independent measurements as detailed in the main text. Panel **b** shows the complete measurement over the full range of the rotation angle α. It is worth noticing that the experimental data without correcting for technical noise do not show number squeezing $\xi^{2N} < 0$ dB around the second minimum of the theory. From the rotation angle dependence of the technical noise we can estimate the main noise sources in the experiment as detailed in Sect. 4.5.1 and appendix C. This figure is originally published in [42]

where the non-linear beamsplitter sequence is depicted which is equivalent to the tomography sequence but for a fixed rotation angle α.

Optimal Number Squeezing

The optimal number squeezing is $\xi_N^2 = -6.9^{+0.8}_{-0.9}$ dB detected under an angle of $\alpha_0 = 16.5°$ where the errors are one standard deviation of the mean over several datasets—in total 634 experimental realizations contribute.[13] We use the six-well one-dimensional optical lattice trap explained in In Sect. 3.1.2 in all experiments on internal spin squeezing, but the barrier separating the individual sites is very high ($V_0 \approx 2\pi \cdot 2.5$ kHz) such that the individual condensates are independent. As detailed in Sect. 4.5 this is an experimental trick to increase the statistics and to estimate the technical noise stemming from single particle effects. Data from different wells contribute to the data shown in Fig. 4.10 and we use all wells with total atom number between 200 and 450, while the mean total atom number over all datasets is 400.

---

[13] In order to obtain better statistics we average the measurements for $\alpha = 16°$ and $\alpha = 17°$.

Subtracting technical noise (see Sect. 4.5 for details) we find an optimum number squeezing of $\xi_N^2 = -8.2^{+0.9}_{-1.2}$ dB close to the theoretical optimum. As this value is cleaned from all known extra noise contributions it is the best estimation of the true variance of the quantum state. The black solid data points in Fig. 4.10 show the technical noise corrected data.[14]

Measuring the Coherence

A rotation by $\alpha = \alpha_0 + 90°$ transforms the most uncertain spin component to the J_z direction. Here we measure increased number fluctuations of $\xi_{N,\max}^2 = 10.3^{+0.3}_{-0.4}$ dB. These fluctuations limit the coherence of the quantum state: Assuming, as a gedanken experiment, we rotate the most uncertain axis to the J_y direction (a rotation by α_0) then $\Delta \hat{J}_y^2 = \frac{N^2}{4} \Delta \varphi^2$. For $N \gg 1, \varphi \ll \pi$ and a gaussian probability distribution $p(\varphi)$ the coherence follows to:

$$\langle \cos(\varphi) \rangle = \frac{\int d\varphi \cos(\varphi) p(\varphi)}{\int d\varphi p(\varphi)} = e^{-\Delta \varphi^2 / 2} \tag{4.14}$$

Noticing that after a unitary rotation by $\alpha = \alpha_0 + 90°$ the maximum number fluctuations $\Delta \varphi^2 = \frac{4 \Delta \hat{J}_z^2}{N^2} = \xi_{N,\max}^2 / N$, measure these phase fluctuations then the coherence is given by:

$$\langle \cos(\varphi) \rangle = e^{-\xi_{N,\max}^2 / 2N} \tag{4.15}$$

Validity of the symmetric two-mode model is crucial here and we tested for it experimentally as described in Sect. 4.6 The coherence follows to $\langle \cos(\varphi) \rangle = 0.986 \pm 0.001$ and we find coherent spin squeezing $\xi_S^2 = \xi_N^2 / \langle \cos(\varphi) \rangle^2$ of:

$$\xi_S^2 = \frac{\xi_N^2}{\langle \cos(\varphi) \rangle^2} = -8.2 \, \text{dB} \tag{4.16}$$

This large amount of coherent spin squeezing allows in principle for a gain of 61% in the phase precision $\Delta \varphi$ of ideal Ramsey type interferometry with respect to the standard quantum limit [43].

Comparison to the One Axis Twisting Theory

The black line in Fig. 4.10 is the theoretical prediction detailed at the end of Sect. 4.2 without any adjustable parameter. The nonlinearity assumed is $\chi = 2\pi \cdot 0.063$ Hz as extracted from the mean field spin dynamics experiment presented in Sect. 4.1.3. The main discrepancy between theory and experimental data is in the squeezed regions. We attribute this difference to a loss of approximately 15%

---

[14] As explained in In Sect. 3.4.1 we always remove the photon shot noise in these experiments.

of the atoms during the total experiment duration of ca. 20 ms which degrades the achievable squeezing [12]. The good agreement with the theory shows that our measurements are well described by the two-mode model and the observed phase dispersion can be fully explained by non-linear one axis twisting dynamics.

4.4 Quantifying Many-Body Entanglement

Coherent spin squeezing is one example where entanglement provides a quantum resource useful to overcome limits set by single particle quantum mechanics [6, 43]. In 2001 Sørensen and Mølmer showed how the fluctuations in one perpendicular spin direction and the coherence of the system can be used to measure many-body entanglement in the system (see also In Sect. 2.3.3) [44]. We are able to detect both quantities and thus we use the *depth of entanglement* measure to quantify entanglement in our system. Fig. 4.11 shows our measurement in context of this quantitative criterion where we use the analytic approximation from Equation (2.24) to plot the theory lines in the figure.

The measured values for coherence $\langle\cos(\varphi)\rangle = 0.986 \pm 0.001$ and number squeezing $\xi_N^2 = -8.2^{+0.9}_{-1.2}$ dB imply entanglement of 170 particles in the sense of the non-separable block size of the many-body density matrix. On a three standard deviation statistical uncertainty level we can exclude less than 80 entangled particles in the system.

4.5 Many Experiments in Parallel: More Than Just Better Statistics

As already mentioned in Sect. 4.3 the Rubidium atoms are trapped in an one dimensional optical lattice. The total atom number is typically 2, 300 distributed over six wells of the lattice and tunneling between the wells is negligible resulting in independent condensates in the individual wells. Figure 4.12 shows the typical total atom number per lattice site. The central wells contain approximately 400atoms,[15] the outer ones 100 to 200 atoms. The local trapping frequencies in each well are $\omega_x = \omega_y = 2\pi \cdot 425$ Hz and $\omega_z = 2\pi \cdot 420$ Hz. This optical lattice configuration has a few important advantages over the single trap configuration. The most obvious one is the increased statistics since we perform six experiments in parallel. But there are two more points worth noticing:

---

[15] Depending on the relative position of dipole trap and optical lattice the central well can contain up to 450 atoms.

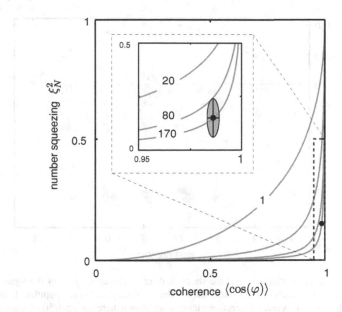

Fig. 4.11 Quantifying the depth of entanglement. This figure shows the measured number fluctuations and coherence in context of the entanglement measure proposed in [44]. The *theory lines* are the analytic approximation of the numerical results from reference [44] underestimating the amount of entangled particles as shown in Fig. 5. The numbers on the different lines give the minimum nonseparable block size of the density matrix. The inset shows our measurement centered at the *line* for 170 entangled particles. The *gray shaded* ellipse is the three standard deviation uncertainty region excluding less than 80 entangled particles on this statistical precision level. This figure is originally published in [42]

Suppressed External Dynamics

The lattice increases the local trap frequency in z-direction from $2\pi \cdot 20$ to $2\pi \cdot 420$ Hz. Compared to the—now in all directions—large trap frequencies the difference in the mean field potentials for atoms in state $|a\rangle$ and $|b\rangle$ is small, such that only small amplitude dynamics are initiated after an abrupt internal state change. The single spatial mode approximation—both modes share the same spatial wavefunction—is therefore valid. Figure 4.13 shows the dynamics of the mean field wavefunction overlap after a $\pi/2$ pulse simulated with the two component Gross-Pitaevskii equation for our parameters.

Technical Noise

Spin readout is performed in a destructive way, meaning a new Bose–Einstein condensate has to be prepared for each single measurement. Since fluctuation measurements require ensemble averaging our results are sensitive to shot-to-shot fluctuations of the experimental parameters. But next to these shot-to-shot variations also

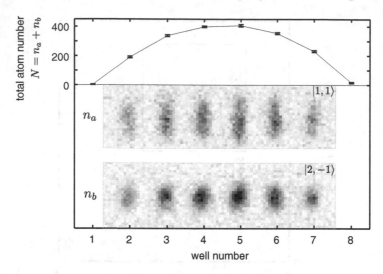

Fig. 4.12 Atom number distribution over the six optical traps. The *upper part* of the figure shows the total atom number in each of the optical traps—the six central wells are populated, well one and eight are empty. *Below* typical absorption pictures are shown, here for a 50/50 mixture of both species. the lengthened shape of the $|1, 1\rangle$ component is due to the longer expansion time before imaging during which the optical lattice is kept on. Extensive details on the imaging procedure can be found in Appendix A

fluctuations of the differential energy shift during the evolution time have to be taken into account. Therefore the one axis twisting Hamiltonian (4.9) becomes:

$$\hat{H} = -\Delta\omega_0(t)\,\hat{J}_z + \chi\,\hat{J}_z^2 \qquad (4.17)$$

The fluctuations $\Delta\omega_0(t)$ result in random rotation frequencies of the spin state around the J_z axis during the evolution time τ. The resulting phase noise in the integrated rotation angle $\Delta\tilde{\phi}(\tau) = \int_0^\tau \Delta\omega_0(t)\mathrm{d}t$ translates into increased fluctuations in the occupation number difference depending on the rotation angle α of the last coupling pulse in the experimental sequence (see Sect. 4.3)

The main contribution to the differential energy shift originates from magnetic field noise. The $|1, 1\rangle$ and $|2, -1\rangle$ hyperfine states share approximately the same linear Zeeman shift, but at $B \approx 9.1$ G the differential Zeeman shift is $\partial\omega_0/\partial B = 2\pi \cdot 10$ Hz/mG requiring for a very high magnetic field stability. We use an active feedback technique and synchronization of the experiment to the power line frequency in order to stabilize the magnetic field to the 100μ G level (for details see Appendix C).

The spin echo pulse mentioned in Sect. 4.3 is used to reduce the low frequency phase noise sensitivity of the system. The normalized spectral sensitivity with and without spin echo pulse is shown in Fig. 4.14. It depends on the total sequence length

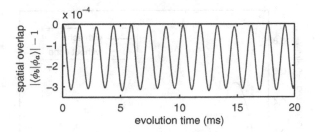

Fig. 4.13 Dynamics of the wavefunction overlap. The figure shows the calculated dynamics in the wavefunction overlap after an abrupt internal state change, e.g., due to a $\pi/2$ pulse. The wavefunction overlap shows negligible dynamics on the 10^{-4} level. The simulation was done for 500 atoms in total in a $\omega_x = \omega_y = 2\pi \cdot 425$ Hz and $\omega_z = 2\pi \cdot 420$ Hz trap and for the scattering length a_{ab} decreased by 10% from its background value

τ and can in principle be modified using further echo pulses—a "bang-bang control" technique [45].[16]

Long Time Coherence Measurement

We perform a Ramsey type coherence measurement and compare the observed visibility of Ramsey fringes versus time for two experiments—with and without echo pulse. The experimental sequence starts with a $\pi/2$ pulse which prepares a coherent spin state on the equator of the Bloch sphere. After a chosen hold time t_{hold} a second $\pi/2$ pulse recombines the two modes for phase sensitive readout. If a spin echo pulse is used, the free evolution is interrupted symmetrically by the π pulse at $t_{hold}/2$. Figure 4.15 shows the measured visibility of the $|1, 1\rangle$ component as a function of hold time t_{hold}. The experiment was done at a magnetic field of $B \gg 9.1$ G away from the Feshbach resonance such that coherent phase spreading due to the non-linear interaction is small.

The observed decay of the visibility can be fitted by a gaussian $\mathcal{V} = e^{-t^2/2\tau_{dec}^2}$ in both cases. We find with $\tau_{dec} = 108$ ms without and $\tau_{dec} = 325$ ms with the spin echo pulse. Low frequency magnetic field fluctuations that represent a non-markovian bath [46] are expected to cause this kind of coherence decay that can be partially cancelled by the spin echo technique. Since spin relaxation loss is not negligible on the experimental timescales the total atom number decreases with a lifetime of a few hundred milliseconds. The data shown in Fig. 4.15 does not take this decay into account explicitly, meaning the visibility is extracted from the observed Ramsey fringes normalized to the total atom number detected at each time t_{hold}.

---

[16] Experimentally the number of echo pulses should be kept minimal, since the coupling pulses introduce additional noise due to fluctuations of pulse phase and power.

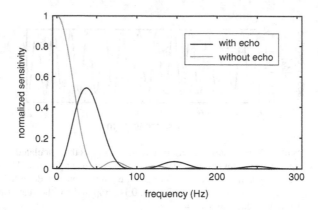

Fig. 4.14 Spectral sensitivity to phase noise. The figure shows the normalized spectral sensitivity of the system to phase noise for a free evolution time of 20 ms. The *black line* shows the altered sensitivity for a spin echo sequence which is normalized to the zero frequency sensitivity limit in the no-echo case (*gray line*). The most important effect of the echo pulse is to remove the zero frequency phase noise sensitivity

4.5.1 Real Time Estimation of Technical Noise

Big technical effort is necessary to minimize the environmental noise and to detect spin squeezing in our system (see Appendix C). However, excess noise is not perfectly cancelled which leads to observed spin fluctuations larger than the intrinsic fluctuations of the quantum state. The six-well trap configuration can be used to monitor this technical noise "real time" and to remove it from the variance measurements. The fluctuations of the measured J_z component of the spin vector can be translated into angular fluctuations in θ, the polar angle on the Bloch sphere. Noise sources acting on the single particle level such as coupling pulse errors or the integrated phase noise due to differential energy shifts cause angular errors that—depending on the experimental sequence—add to the fluctuations of θ. Due to the single particle nature of these effects the contribution to the observed fluctuations $\Delta \hat{J}_{z,\text{tech}}^2 = \beta^2 J^2$ is quadratic in the total spin length $J = N/2$. Here β^2 indicates the angular fluctuations in polar direction at the time of measurement stemming from technical noise.

Performing six experiments in parallel offers the possibility to check the dependence of the observed occupation number fluctuations on the total atom number for each measured dataset—"real time"—such that $\Delta \hat{J}_{z,\text{tech}}^2$ can be subtracted accurately. We bin the individual wells in all possible combinations and bin sizes and calculate the fluctuations for each binning. This procedure allows to calculate the number squeezing ξ_N^2 for different total atom numbers N. Figure 4.16 illustrates this procedure and shows the obtained correlation of the number squeezing and the total atom number for one exemplary data set. The correlation is due to the technical

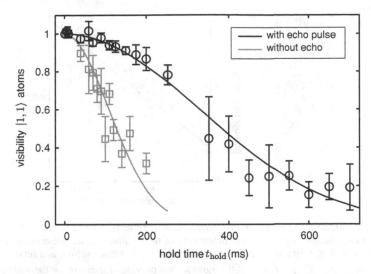

Fig. 4.15 Ramsey type coherence measurement. The figure shows the visibility of a Ramsey fringe extracted from the atoms in the $|1, 1\rangle$ state measured versus hold time t_{hold}. Due to a detection problem at the time the measurement was done the information of the second spin state $|2, -1\rangle$ can not be used. However the purpose of the figure is to estimate the effect of the spin echo pulse which is nicely shown by the data. The coherence time can be significantly enhanced by the π echo, a strong indication for low frequency phase noise present in our experimental setup

noise which affects all—otherwise independent—condensates in the same way.[17] We extract the slope β^2 and remove the technical variance $\Delta \hat{J}^2_{z,\text{tech}}(N)$ from the variance measured in each individual well with total atom number N.

In the noise tomography experiment described in Sect. 4.3 the correction is small for small rotation angles but without noise removal no number squeezing is detected after a $\alpha = \alpha_0 + 180°$ rotation. This indicates pulse power fluctuations or shot to shot magnetic field drifts to which this rotation close to $180°$ is most sensitive (see Appendix C).[18] For $\alpha = 90°$ the measurement is maximally sensitive to longitudinal phase noise $\Delta\tilde{\phi}$ originating from differential energy shifts between the two modes. Figure 4.17 shows the normalized experimental noise versus rotation angle detected in the tomography experiment.

---

[17] The spacing between the wells is only 5.7μ m such that magnetic field fluctuations and the electromagnetic radiation fields for the coupling are homogeneous over the whole system.

[18] Fluctuation measurements on a coherent spin state after a $7\pi/2$ pulse in a Rabi cycle still show shot noise limited noise characteristics. This indicates negligible pulse power fluctuations in our experiment and suggests again that shot-to-shot magnetic field fluctuations are the main noise source.

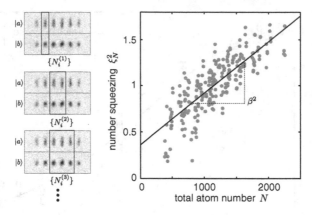

Fig. 4.16 Technical noise estimation. Performing six experiments in parallel, spatially spaced by a few micrometer, allows for "real time" monitoring of the technical noise. Noise measurement is done by binning of the data for all different combinations of k wells resulting in a set of different total atom numbers $\{N_i^{(k)}\}$ (*left part* of the figure). This provides a measure for the scaling of the normalized occupation number fluctuations ξ_N^2 with total atom number N. The linear slope β^2 is extracted for each dataset and the measured fluctuations in each individual well are corrected for the technical noise. The graph on the *right* of the figure shows the result of this procedure for one exemplary dataset

4.6 Heisenberg Minimal Uncertainty Product and Validity of the Symmetric Two-Mode Model

Validity of the Symmetric Two-Mode Model

Some arguments given in the previous sections for example the calculation of the coherence from the maximum anti-squeezing $\xi_{N,\max}^2$ in Sect. 4.3 require the validity of the symmetric two-mode model. We test this assumption by comparing the two-mode coherence $\langle \cos(\varphi) \rangle_{2m} = e^{-\xi_{N,\max}^2/2N}$ with the coherence $\langle \cos(\varphi) \rangle = \mathcal{V}$ measured via the visibility $\mathcal{V}$ in a Ramsey experiment. The Ramsey method reveals the actual coherence taking all deteriorating effects into account. In order to perform the Ramsey measurement we prepare a spin squeezed state by the one axis twisting sequence detailed in Sect. 4.3 The last coupling pulse which was used to rotate the quantum state around its center in the previous experiment is now replaced by a $\pi/2$ pulse whose phase φ is changed over the full $[0, 2\pi]$ interval.

We use the data from the central four wells obtained in 847 experimental repetitions resulting in 3, 388 data points in total. Figure 4.18 shows the obtained Ramsey fringe where the data points represent the normalized population imbalance n/N measured in the different experiments and the black solid line is a sinusoidal fit to the data. The fit reveals a visibility $\mathcal{V} = 1.00 \pm 0.02$, confirming the two-mode model from which a visibility of $\langle \cos(\varphi) \rangle_{2m} = 0.986 \pm 0.001$ was predicted. Since the coherence deduced from the maximum anti-squeezing is more accurate, we use

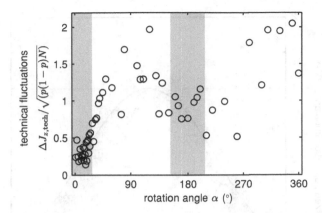

Fig. 4.17 Rotation dependence of the technical noise. The figure shows the normalized technical noise contribution in the tomography experiment. The noise shown here is the difference between the *open gray* and the *solid black* data presented in Fig. 10. Within the first 180° of rotation the noise is maximal for angles around 90° identifying longitudinal phase noise $\Delta\tilde{\phi}$ as the main noise contribution. The difference in the noise level between the two *gray shaded areas* can be explained by shot-to-shot magnetic field fluctuations, which result in fluctuations of the coupling pulse rotation axis in polar direction (see Appendix C)

this value for the various calculations presented, e.g. coherent spin squeezing, number of entangled particles and the Heisenberg uncertainty product.

Heisenberg Uncertainty Product

As discussed in In Sect. 2.2 the Heisenberg minimal uncertainty product for the spin operators is

$$\Delta_H^2 = \frac{4\Delta \hat{J}_z^2 \Delta \hat{J}_y^2}{\langle \hat{J}_x^2 \rangle} \geq 1 \tag{4.18}$$

which can be expressed in the measured quantities ξ_N^2 and $\langle \cos\varphi \rangle$:

$$\Delta_H^2 = \frac{\xi_{N,\min}^2 \xi_{N,\max}^2}{\langle \cos(\varphi)^2 \rangle} \tag{4.19}$$

Experimentally we find

$$\Delta_H^2 = 1.65 \pm 0.35 \tag{4.20}$$

which is only slightly larger than the value of $\Delta_H^2 = 1.01$ predicted by the two-mode theory. The discrepancy between the two numbers can be explained from the difference of the best measured number squeezing of $\xi_{N,\min}^2 = -8.2$ dB to the value $\xi_{N,\min,\text{Ueda}}^2 = -10.3$ dB which is the theoretical prediction without particle loss. The

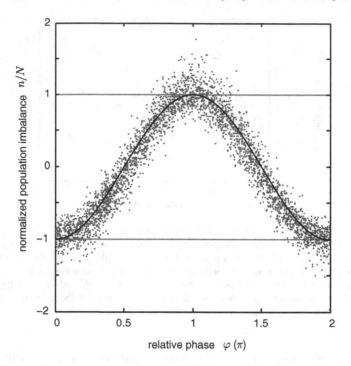

Fig. 4.18 Ramsey fringe using a spin squeezed state. The figure shows a Ramsey fringe obtained for a spin squeezed state prepared using the 18 ms one axis twisting sequence explained in Section 4.3 From the sinusoidal fit we extract a visibility of $\mathcal{V} = 1.00 \pm 0.02$ in good agreement with 0.986 ± 0.001, the value deduced assuming the symmetric two-mode model

ratio of these two numbers is 1.62, identifying particle loss as the main deteriorating effect in this measurement.

4.7 Non-linear Atom Interferometer Beats "Classical" Precision Limit

For a long time interaction among particles has been regarded as a drawback for atom interferometry [47, 48]. Recent theoretical work however revealed that-in principle-one axis twisting dynamics does not spoil interferometric precision and even more that it can lead to interferometry close to the ultimate Heisenberg limit [34].

We experimentally realize a novel non-linear atom interferometer and show interferometric precision beyond the standard quantum limit. The interferometric scheme is related to a standard Ramsey interferometric sequence, where the accumulated phase φ between two modes is measured (see In Sect. 2.4). The phase of interest is accumulated within a time τ which is bounded between two $\pi/2$ coupling pulses.

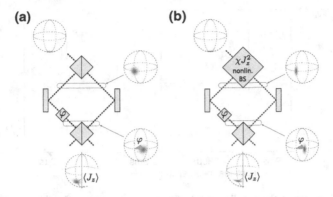

Fig. 4.19 Optical analog to our non-linear interferometer. Panel **a** shows the analog of optical Mach-Zehnder interferometry to a standard Ramsey sequence for an atomic interferometer based on internal states of the atoms. The analog to a beamsplitter in optics is a $\pi/2$ pulse and the acquired phase φ, the longitudinal angle on the *Bloch sphere*, translates into an population imbalance at the output. **b,** Atom interferometry beyond the standard quantum limit can be realized by the replacement of the first beamsplitter with a non-linear one. The non-linear beamsplitter produces an entangled— phase squeezed state—at its output where the reduced variance in longitudinal direction translates into reduced noise in the population imbalance J_z at the output

These two pulses are the analog to beamspitters in optical Mach-Zehnder interferometry (see Fig. 4.19a) [49], where the first beamsplitter creates a coherent superposition of the two modes $|a\rangle$ and $|b\rangle$ while the last pulse is necessary to translate the acquired phase into a observable population difference.

A non-linear Beamsplitter for Bose–Einstein Condensates

In our non-linear atom interferometer the first beamsplitter is replaced by a *non-linear beamsplitter* (Fig. 4.19b). At its output a coherent spin squeezed state appears which propagates for a time τ until a standard linear beamsplitter—a $\pi/2$ pulse—couples the two modes before readout. As described in In Sect. 2.4 coherent spin squeezed states allow for interferometric precision beyond the standard quantum limit.

The realization of the non-linear beamsplitter is closely related to the noise tomography experiment presented in Sect. 4.3 and its implementation is detailed in Fig. 4.20. We use the same experimental parameters as for the tomography experiment, in particular the magnetic field is constant at $B = 9.13$ G and the twisting time is 18 ms symmetrically split by a spin echo pulse. The angle of the last rotation pulse around the center of the quantum state is chosen to $\alpha = \alpha_0 + 90°$, such that the spin direction with minimal fluctuations is in J_y direction—a *phase squeezed state* is prepared. We choose a short interferometric evolution time of $\tau_{evo} = 2\ \mu$s in order to avoid magnetic field fluctuations to spoil our measurement.

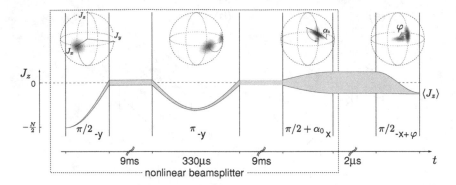

Fig. 4.20 Schematic of the non-linear interferometric sequence. The figure summarizes the non-linear beamsplitter and the following interferometric sequence graphically. The action of each coupling pulse is shown on the Bloch sphere and the evolution of the J_z component of the spin and its variance is indicated by the gray shaded curve and its width. The rotation axis and angle of each pulse and the experimental durations of the different intervals are given below. The non-linear beamsplitter sequence (dashed area) is a special case of the noise tomography sequence described in Section 4.3 with the last rotation chosen to $\alpha = \alpha_0 + 90°$

Performance of the Non-linear Interferometer

For a straightforward and very intuitive characterization of the performance of the non-linear interferometer we carry out repeated measurements around the working point of the interferometer and calculate the phase precision. We vary the mean acquired phase $\langle \varphi \rangle$ across an interval $[-16°, 16°]$ and detect the mean population imbalance $\langle \hat{J}_z \rangle$ and its variance $\Delta \hat{J}_z^2$ (Fig. 4.21). This allows us to draw an uncertainty band whose horizontal width gives the phase measurement precision $\Delta \varphi$. In order to compare the performance of the non-linear interferometer to a standard linear Ramsey scheme, we replace the non-linear beamsplitter by a single $\pi/2$ pulse and repeat the same measurement. Since approximately 15% of the atoms are lost during the non-linear sequence (which takes approximately 18.5 ms longer than the linear one) we experimentally adjust the atom number in the linear case in order to have a similar number at the time of detection. We find an increase of phase precision of 31% when using the non-linear interferometer as compared to the linear one, however due to the presence of classical noise—mainly magnetic field noise and readout photon shot noise—this number alone does not imply improvement beyond the standard quantum limit. We stress that no noise correction is done on the data presented in this section–the shown data points represent the raw data after filtering for rare outliers (see In Sect. 3.4.1).

In order to claim measurement precision above the standard quantum limit it is important to compare the results to the precision of an *ideal* linear interferometer that does not suffer from any technical noise. Knowledge of the coherence of the quantum state is crucial for this comparison and we measure the visibility $\mathcal{V}$ of a Ramsey fringe obtained when scanning the phase φ in the interferometer over the full

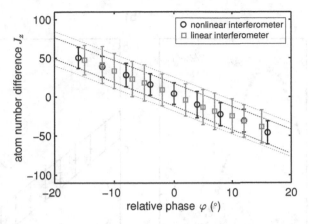

Fig. 4.21 Comparison of the linear and non-linear interferometer. We directly compare the performance of the linear and non-linear interferometer. An offset has been added to center the results of both measurements relative to each other. The *black (gray) dashed lines* are a fit through the upper and lower ends of the two standard deviation error bars for the non-linear (linear) interferometer. The *horizontal width* of the included areas measure the phase precision. We find 31% increased phase precision for the non-linear interferometer as compared to the linear one

interval $[0, 2\pi]$. We find $\mathcal{V} = 0.98 \pm 0.02$ for the linear and $\mathcal{V} = 0.92 \pm 0.02$ for the non-linear interferometer (Fig. 4.22a). The signal for an ideal linear interferometer is inferred by calculating the expected fluctuations for a binomial distribution $\Delta J_{z,\text{sn}} = \sqrt{p(1 - p)N}$—the shot noise level—for the data measured at each mean value of the relative phase $\langle \varphi \rangle$.[19] This reveals error bars whose upper and lower ends are linearly fitted and the slope m of the fits is corrected by $m/\mathcal{V}$ in order to take the decreased coherence into account. Figure 4.22b shows the data of Fig. 4.21 and additionally the uncertainty regions expected for an ideal linear interferometer. Even though technical imperfections deteriorate the precision of the non-linear interferometer we find

$$\frac{\Delta \varphi_{\text{nl}}}{\Delta \varphi_{\text{l}}} = 0.85 \tag{4.21}$$

where $\Delta \varphi_{\text{nl(l)}}$ is the phase error of the non-linear (ideal linear) interferometer, implying 15% increased phase sensitivity beyond the standard quantum limit.

The linear interferometer measurements show 24% larger phase noise than expected for an ideal interferometer highlighting the effect of excess technical noise. In order to estimate the amount of this noise in the non-linear measurement we calculate the mean number squeezing over all datasets for the non-linear interferometer which are shown in Fig. 4.22b and find $\xi_N^2 = -2.1$ dB. Subtracting photon shot noise and technical noise (see Sects. 3.4.1 and 4.5) we calculate $\xi_N^2 = -4.3$ dB. This value still differs from the best measured number squeezing at the output of the

---

[19] As a reminder, $p = \langle n_a/N \rangle$ is the probability for an atom to be found in mode $|a\rangle$.

(a)   **(b)**

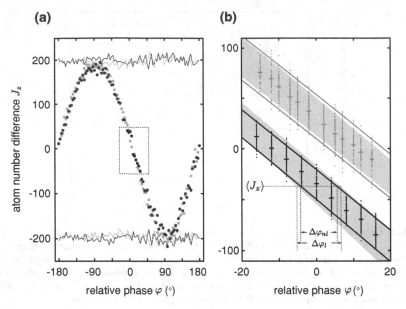

Fig. 4.22 Interferometry beyond the standard quantum limit. Panel **a** shows a Ramsey fringe recorded at the output of the linear (*gray*) and non-linear interferometer (*black*). The *solid lines* give the total number of atoms $N/2$ as a reference. We find a coherence of 0.98 ± 0.02 in the linear and 0.92 ± 0.02 in the non-linear case. **b,** Comparison of the interferometer performance to an ideal linear Ramsey interferometer (*gray shaded areas*) reveals an increase in phase precision by 15% when using the non-linear atom interferometer (*black*). The individual data points are shown without any noise subtraction and the *solid lines* are linear fits through the *lower* and *upper* ends of two standard deviation error bars. Our implementation of the linear Ramsey scheme performs 24% worse than the ideal one (*gray*) highlighting the effect of technical induced fluctuations. An offset has been added to separate the two measurements for clarity. This figure is originally published in [42]

non-linear beamsplitter ($\xi_N^2 = -8.2$ dB). This difference is due to the long measurement duration for the full interferometer dataset which is in the order of 24 h. Slow magnetic field drifts (on the order of 5 mG per day) become important since the nonlinearity χ changes and the performance of the non-linear interferometer degrades.

The best detected coherent spin squeezing the output of the the non-linear beamsplitter of $\xi_S^2 = -8.2$ dB (see Sect. 4.3) allows in principle for a phase precision gain of 61% compared to the standard quantum limit. In a possible future experimental setup magnetic field noise and read out noise due to the detection process have to be reduced in order to make use of the full precision increase. Nevertheless our experiment shows the feasibility of non-linear atom interferometry based on interacting atoms and together with novel readout techniques it might lead to real life interferometric sensors operating beyond the standard quantum limit.

References

1. Leggett A (2001) Bose–Einstein condensation in the alkali gases. Some fundamental concepts. Rev Mod Phys 73:307–356
2. Pezzé L, Smerzi A, Berman GP, Bishop AR, Collins LA (2006) Nonlinear beam splitter in Bose–Einstein-condensate interferometers. Phys Rev A 74:033610
3. Gati R (2007) Bose–Einstein condensates in a single double well potential. Ph.D. thesis, University of Heidelberg
4. Santarelli G et al (1999) Quantum projection noise in an atomic fountain. A high stability cesium frequency standard. Phys Rev Lett 82:4619–4622
5. Vandersypen LMK, Chuang IL (2005) NMR techniques for quantum control and computation. Rev Mod Phys. 76:1037–1069
6. Sørensen AS, Duan L, Cirac J, Zoller P (2001) Many-particle entanglement with Bose–Einstein condensates. Nature 409:63–6
7. Kaufman AM et al (2009) Radio-frequency dressing of multiple Feshbach resonances. Phys Rev A 80:050701
8. Widera A et al (2004) Entanglement interferometry for precision measurement of atomic scattering properties. Phys Rev Lett 92:160406
9. Erhard M, Schmaljohann H, Kronjäger J, Bongs K, Sengstock K (2004) Measurement of a mixed-spin-channel Feshbach resonance in ^{87} Rb. Phys Rev A 69:032705
10. Steel M, Collett M (1998) Quantum state of two trapped Bose–Einstein condensates with a Josephson coupling. Phys Rev A 57:2920–2930
11. Sinatra A, Castin Y (2000) Binary mixtures of Bose–Einstein condensates. Phase dynamics and spatial dynamics. Eur Phys J D 8:319
12. Li Y, Treutlein P, Reichel J, Sinatra A (2009) Spin squeezing in a bimodal condensate. spatial dynamics and particle losses. Eur Phys J B 68:365–381
13. Poulsen UV, Mølmer K (2001) Positive-P simulations of spin squeezing in a two-component Bose condensate. Phys Rev A 64:013616
14. Hall DS, Matthews MR, Wieman CE, Cornell EA (1998) Measurements of relative phase in two-component Bose–Einstein condensates. Phys Rev Lett 81:1543–1546
15. Matthews MR et al (1998) Dynamical response of a Bose–Einstein condensate to a discontinuous change in internal state. Phys Rev Lett 81:243–247
16. Treutlein P, Hommelhoff P, Steinmetz T, Hänsch TW, Reichel J (2004) Coherence in microchip traps. Phys Rev Lett 92:203005
17. Cohen-Tannoudji C, Dupont-Roc J, Grynberg G (1992) Atom-photon interactions: basic processes and applications. Wiley, New York
18. Milburn G, Corney J, Wright E, Walls D (1997) Quantum dynamics of an atomic Bose–Einstein condensate in a double-well potential. Phys Rev A 55:4318–4324
19. Javanainen J, Ivanov M (1999) Splitting a trap containing a Bose–Einstein condensate. Atom number fluctuations. Phys Rev A 60:2351–2359
20. Bohi P et al (2009) Coherent manipulation of Bose–Einstein condensates with state-dependent microwave potentials on an atom chip. Nat Phys 5:592–597
21. Chin C, Grimm R, Julienne P, Tiesinga E (2010) Feshbach resonances in ultracold gases. Rev Mod Phys 82:1225
22. Hall DS, Matthews MR, Ensher JR, Wieman CE, Cornell EA (1998) Dynamics of component separation in a binary mixture of Bose–Einstein condensates. Phys Rev Lett 81:1539–1542
23. Timmermans E (1998) Phase separation of Bose–Einstein condensates. Phys Rev Lett 81:5718–5721
24. Li Y, Castin Y, Sinatra A (2008) Optimum spin squeezing in Bose–Einstein condensates with particle losses. Phys Rev Lett 100:210401
25. Raghavan S, Smerzi A, Fantoni S, Shenoy SR (1999) Coherent oscillations between two weakly coupled Bose–Einstein condensates. Josephson effects, π oscillations, and macroscopic quantum self-trapping. Phys Rev A 59:620–633

26. Widera A et al (2005) Coherent collisional spin dynamics in optical lattices. Phys Rev Lett 95:190405
27. Kitagawa M, Ueda M (1993) Squeezed spin states. Phys Rev A 47:5138–5143
28. Bachor HA, Ralph TC (2004) A guide to experiments in quantum optics. Wiley, New York
29. Sinatra A, Castin Y (1998) Phase dynamics of Bose–Einstein condensates. Losses versus revivals. Eur Phys J D 4:247–260
30. Wright EM, Walls DF, Garrison JC (1996) Collapses and revivals of Bose–Einstein condensates formed in small atomic samples. Phys Rev Lett 77:2158–2161
31. Yurke B, Stoler D (1986) Generating quantum mechanical superpositions of macroscopically distinguishable states via amplitude dispersion. Phys Rev Lett 57:13–16
32. Greiner M, Mandel O, Hänsch TW, Bloch I (2002) Collapse and revival of the matter wave field of a Bose–Einstein condensate. Nature 419:51–54
33. Ferrini G, Minguzzi A, Hekking FWJ (2008) Number squeezing quantum fluctuations and oscillations in mesoscopic Bose Josephson junctions. Phys Rev (A) 78:023606
34. Pezzé L, Smerzi A (2009) Entanglement nonlinear dynamics and the Heisenberg limit. Phys Rev Lett 102:100401
35. Pezzé L, Collins LA, Smerzi A, Berman GP, Bishop AR (2005) Sub-shot-noise phase sensitivity with a Bose–Einstein condensate Mach-Zehnder interferometer. Phys Rev (A) 72:043612
36. Appel J et al (2009) Mesoscopic atomic entanglement for precision measurements beyond the standard quantum limit. Proc Natl Acad Sci U S A. 106:10960–10965
37. Fernholz T (2008) Spin squeezing of atomic ensembles via nuclear-electronic spin entanglement. Phys Rev Lett 101:073601
38. Ramsey NF (1950) A molecular beam resonance method with separated oscillating fields. Phys Rev 78:695–699
39. Ramsey NF (1949) A new molecular beam resonance method. Phys Rev 76:996
40. Albiez M (2005) Observation of nonlinear tunneling of a Bose–Einstein condensate in a single Josephson junction. Ph.D. thesis, University of Heidelberg
41. Metcalf H, Vander Straten P (1999) Laser cooling and trapping. Springer, New York
42. Gross C, Zibold T, Nicklas E, Estève J, Oberthaler MK (2010) Nonlinear atom interferometer surpasses classical precision limit. Nature 464:1165–1169
43. Wineland D, Bollinger J, Itano W, Heinzen D (1994) Squeezed atomic states and projection noise in spectroscopy. Phys Rev A 50:67–88
44. Sørensen AS, Mølmer K (2001) Entanglement and extreme spin squeezing. Phys Rev Lett 86:4431–4434
45. Viola L, Lloyd S (1998) Dynamical suppression of decoherence in two-state quantum systems. Phys Rev (A) 58:2733–2744
46. Ferrini G, Spehner D, Minguzzi A, Hekking FWJ (2010) Anomalous decoherence rate of macroscopic superpositions in Bose Josephson junctions. Phys Rev A 82:033621
47. Search CP, Meystre P (2003) Noise limits in matter-wave interferometry using degenerate quantum gases. Phys Rev (A) 67:061601
48. Scully MO, Dowling JP (1993) Quantum-noise limits to matter-wave interferometry. Phys Rev (A) 48:3186–3190
49. Cronin AD, Schmiedmayer J, Pritchard DE (2009) Optics and interferometry with atoms and molecules. Rev Mod Phys 81:1051

Chapter 5
Outlook

The obtained results on spin squeezing in a Bose–Einstein condensate on both external and internal degrees of freedom open up a very promising road toward applied quantum atom optics. A non-linear atom interferometer with absolute precision competitive to the limit of todays best linear interferometers becomes within reach for real life measurements.

For an atom interferometer based on external degrees of freedom a more specialized trap design with higher spatial stability, maybe involving non-harmonic potentials, might be used to overcome the limitations in stability and tunability. The realization of a beam spitter analog among the external modes is a defined goal in this context. Entropy control in splitting direction is another issue here. The production of larger number squeezing requires lower entropy in the Josephson many-body modes which might be realizable with a trap design such that the trap frequency in splitting direction is larger than the thermal energy scale. This transversal splitting has been realized in atom chip interferometers [1], where a first evidence of number squeezing has been observed [2]. State dependent potentials on atom chips have been used for the generation of spin squeezed states for which spin squeezing has been directly observed [3].

The extraordinary experimental control in the experiments based on internal atomic states allowed for the realization of a prototypal non-linear interferometer. Our result shows the validity of the simple two-mode model for this system and is in agreement with the predictions of references [4, 5]. The authors also calculated the parameters for optimal spin squeezing including particle loss for a large number of atoms ($N = \mathcal{O}\,(10^5)$). Given this number of atoms, state of the art microwave technology together with well designed magnetic shielding might already allow for non-linear precision measurements with ultracold atoms that compete with the best available linear measurements.

We developed the experimental technology to detect and control a Bose–Einstein condensate in two different two-mode systems with very high precision. These experiments can in principle be combined resulting in effectively four modes among which cross interaction and coupling is controllable. The two additional degrees of freedom

C. Groß, *Spin Squeezing and Non-linear Atom Interferometry with Bose–Einstein Condensates*, Springer Theses, DOI: 10.1007/978-3-642-25637-0_5, © Springer-Verlag Berlin Heidelberg 2012

allow for deeper exploration of different types of many-body entanglement. One example is Einstein–Podolsky–Rosen type continuous variable entanglement [6, 7] which, for massive particles, has been detected in vapor cell experiments [7]. However the violation of the continuous variable Einstein–Podolsky–Rosen criterion in this experiment was not strong enough to violate the generalized Einstein–Podolsky–Rosen paradox, which tests for local realism for all 'elements of reality' [7]. Experiments in the four mode Bose–Einstein condensate might overcome this limitation and allow for the realization of a Einstein–Podolsky–Rosen paradox in a macroscopic system consisting of a few hundred atoms [9, 10].

References

1. Schumm T et al (2005) Matter-wave interferometry in a double well on an atom chip. Nat Phys 1:57–62
2. Jo G et al (2007) Long phase coherence time and number squeezing of two Bose–Einstein condensates on an atom chip. Phys Rev Lett 98:030407
3. Riedel MF, Böhi P, Li Y, Hänsch TW, Sinatra A, Treutlein P (2010) Atom-chip-based generation of entanglement for quantum metrology. Nature 464:1170
4. Li Y, Castin Y, Sinatra A (2008) Optimum spin squeezing in Bose–Einstein condensates with particle losses. Phys Rev Lett 100:210401
5. Li Y, Treutlein P, Reichel J, Sinatra A (2009) Spin squeezing in a bimodal condensate: spatial dynamics and particle losses. Eur Phys J B68:365–381
6. Einstein A, Podolsky B, Rosen N et al (1935) Can quantum-mechanical description of physical reality be considered complete?. Phys Rev 47:777–780
7. Reid MD et al (2008) The Einstein–Podolsky–Rosen paradox: from concepts to applications. Rev Mod Phys 81: 1727–1751
8. Julsgaard B, Kozhekin A, Polzik ES (2001) Experimental long-lived entanglement of two macroscopic objects. Nature 413:400–403
9. Bar-Gill N, Gross C, Mazets I, Oberthaler M, Kurizki G (2011) Einstein–Podolsky– Rosen correlations of ultracold atomic gases. Phys Rev Lett 106:120404
10. He QY, Reid MD, Vaughan TG, Gross C, Oberthaler M, Drummond PD (2011) Einstein–Podolsky–Rosen entanglement strategies in two–well Bose–Einstein condensates. Phys Rev Lett 106:120405

Appendix A
Precision Absorption Imaging of Ultracold Atoms

A.1 Hardware and Alignment of the Imaging System

A high resolution absorption imaging system compatible with a typical experiment on Bose–Einstein condensation was developed in our group [1, 2]. We briefly describe the crucial points of the setup and alignment of the imaging system here. For details on the custom made optics we refer the reader to references [1] and [2] and for basics on absorption imaging to [3].

Hardware

The most important part of the imaging system is an infinite conjugate objective featuring a numerical aperture of 0.45. Three custom made lenses, anti reflection coated for 780 nm are contained in this objective. Layout and design of the optics was done by Carl Zeiss Laser Systems, the housing was built by the mechanical workshop of the Kirchhoff Institute for Physics and the mounting was done in our group. At the time of imaging the distance of the objective to the glass cell is approximately 1 mm, blocking all optical access to the experimental chamber from this side. Therefore the objective is mounted on a step motor allowing for 110 mm travel with a repeatable position precision of $1 \mu m$,[1] such that it can be moved away from the glass cell during the MOT phase of the experimental sequence.

The image of the atoms is focused onto a CCD chip using a standard infinite conjugate achromat with a focal length of 1 m. We have chosen a magnification of 30.96, such that the resolution of the system is not limited by the pixel size ($13\mu m$) of our CCD camera. We use a back illuminated deep depletion CCD camera[2] with a quantum efficiency of ca. 93% at a wavelength of $\lambda = 780$ nm. The vacuum

---

[1] MICOS Linear Stage LS-110.
[2] Princeton Instruments, PIXIS: 1024BR.

C. Groß, *Spin Squeezing and Non-linear Atom Interferometry with Bose–Einstein Condensates*, Springer Theses, DOI: 10.1007/978-3-642-25637-0, © Springer-Verlag Berlin Heidelberg 2012

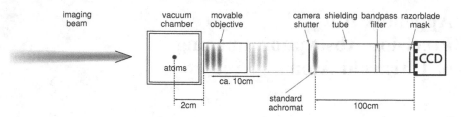

Fig. A.1 Setup of the imaging system. A large numerical aperture movable objective allows for high resolution absorption imaging of the ultracold atoms while it preserves maximal optical access at the laser cooling phase of the experimental cycle. The shielding tube is necessary to minimize air motion in the imaging path and to decouple the CCD camera from the mechanical shutter. Both, objective and CCD camera are mounted mechanical stable such that shot noise limited imaging is achieved. The total length of the setup is approximately 1.5 m set by the required magnification and the dimensions of the step motor that carries the objective

window of the camera is parallel (non-wedged) and double sided anti reflection coated. Two narrow bandpass filters[3] with the transmission centered around $\lambda = 780$ nm are placed before the CCD camera since the objective has to be shielded from laser light originating from the optical lattice beams ($\lambda = 843$ nm) and one of the dipole trap beams ($\lambda = 1064$ nm). We measure 72% quantum efficiency of the whole system including all optics and the uncoated experimental chamber. The numbers given in this paragraph represent the current setup of the imaging system (Fig. A.1) which was used for the squeezing experiments based on internal degrees of freedom. In the first (external squeezing) experiment we used a different CCD camera[4] which has a much lower quantum efficiency[5] and smaller pixel size of 6.45μm. Here the achromat focussing the image onto the CCD chip was chosen such that the resulting magnification was 11.2.

As an estimate for the resolution of our imaging system we measure its point spread function by in situ imaging of a small Bose–Einstein condensate and we find a width (gaussian standard deviation) of ca. 700 nm consistent with the diffraction limit.

Alignment

Correct alignment of the objective with respect to the glass cell of the experimental chamber is crucial to minimize optical aberration. Experimentally most challenging is to obtain parallelism between the glass cell and the principal planes of the objective lenses. Careful mounting and high precision manufacturing of the objective ensures correct relative alignment of all three lenses and their parallelism to the front surface of the objective mount. In order to align the objective to the glass cell a glass plate is glued to this front surface. A reference

---

[3] Semrock BrightLine HC 780/12.

[4] QImaging, Retiga Exi.

[5] The total quantum efficiency of the old setup was approximately 20%.

laser beam is set up such that it hits the position of the atomic cloud and such that it is perpendicular to the glass cell. The remaining challenge is to align the reflected spots from the glass cell and from the glass plate glued to the objective.

The imaging beam is tilted by an angle of approximately 5° to the normal of the glass cell in order to avoid etaloning between the various parallel glass surfaces.

A.2 The Imaging Sequence

In the external squeezing experiment only the $|F = 2, m_F = 2\rangle$ state of the ^{87} Rubidium atoms has to be detected and the imaging sequence is straightforward. Absorption imaging requires at least two pictures, one containing the absorption signal from the atoms and a second one—the reference picture—measuring the light intensity at the position of the atoms. We take these two images temporally spaced by approximately 800 ms. A small offset magnetic field is kept on during the detection and the imaging beam is polarized in order to drive the $|F = 2, m_F = 2\rangle \leftrightarrow |F' = 3, m_{F'} = 3\rangle$ cycling transition.

The squeezing experiment based on the $|F = 1, m_F = 1\rangle$ and $|F = 2, m_F = -1\rangle$ hyperfine states requires a more complicated imaging sequence since the two states have to be distinguishable on the pictures. The experiments are done at a high magnetic field of $B \approx 9$ G, but for imaging we ramp the field down to a value close to zero within 3 ms before the first image is taken. This image contains the absorption information of the $|F = 2, m_F = -1\rangle$ atoms. We use the fast frame transfer mode of our CCD camera to move the first image to the masked area of the chip[6] such that we can take the second picture 780μs later. During the shifting we shine a resonant laser beam which does not enter the imaging objective to remove the already imaged $|F = 2\rangle$ atoms. A few tens of microseconds before the second picture a re-pumping laser is switched on to transfer the population in the $|F = 1, m_F = 1\rangle$ to the $|F = 2\rangle$ hyperfine manifold and the imaging laser's absorption driving the $|F = 2\rangle \leftrightarrow |F' = 3\rangle$ is measured. The $|F = 2\rangle$ atoms are again removed from the field of view of the camera and a reference picture is taken another 780μs later.

In all experiments the optical dipole trap is kept on until 500μs before the first image. From the time of switch off the atomic cloud expands which is necessary in order to avoid non-linear effects spoiling the imaging accuracy (see Sect. A.3.2). The imaging pulse duration is chosen between 5μs and 25μs where transversal blurring limits its duration depending on the number of atoms in the trap.

---

[6] 4/5th of the CCD chip are masked by a razor blade (see Fig. A.1).

A.3 Calibration of the Imaging System

Our experiments are based on atom number fluctuation measurements which requires a accurate calibration of the total atom number and even more the linearity of the imaging system.

A.3.1 Atom Number Calculation

In high intensity absorption imaging [4] the full Beer-Lambert absorption formula is necessary to calculate the atomic column density $n_{i,j}$ from the counts per pixel $N_{\gamma,\text{pic},i,j}$ on the picture containing the absorption information and on the reference picture $N_{\gamma,\text{ref},i,j}$ (i,j indexes the CCD camera pixels but it is mostly omitted below to ensure better readability). Not only the optical density $O_d = \ln(N_{\gamma,\text{ref}}/N_{\gamma,\text{pic}})$ but also the difference in the counts $\Delta = (N_{\gamma,\text{ref}} - N_{\gamma,\text{pic}})$ contributes to the signal:

$$n = \frac{d^2}{\sigma_0}(\frac{1}{c_{cg}}O_d + \frac{c_{ccd}c_{\text{gpe}}}{\tau}\Delta) \tag{A.1}$$

Here d is the linear extension of the CCD pixel taking the magnification into account, $\sigma_0 = 3\lambda^2/2\pi$ the resonant cross section, c_{cg} the Clebsch-Gordan coefficient, c_{gpe} a correction factor obtained from a comparison to simulations as explained below and τ is the imaging pulse duration. The factor $c_{ccd} = \hbar\omega/(d^2\eta Q I_{sat})$ contributes to the linear part of the formula where Q is the total quantum efficiency and η the gain factor of the camera. ω is the light angular frequency and $I_{sat} = 1.67$ mW/cm^2 the saturation energy of the transition.

The total quantum efficiency Q can be measured using a well calibrated power meter and the factor η is determined by measuring the noise features of the CCD camera.

Noise Features of the CCD Camera

We measure the noise curve of the camera over the full dynamic range using an incoherent light source (e.g. a LED), taking typically more than 10,000 measurements per mean camera count. Photons hitting the active region of the CCD are assumed to be uncorrelated featuring shot noise limited noise characteristics. The factor $\eta = 1.025$ follows from the slope of the CCD's noise curve via a linear fit in the working region where the camera noise is not read out noise dominated. For the PIXIS camera the slope of a linear fit is consistent with the value of the linear parameter of a second order polynomial fit. Figure A.2 shows the measured noise curve of the PIXIS camera including a second order fit

$$\Delta N_\gamma^2 = p_1\langle N_\gamma\rangle^2 + p_2\langle N_\gamma\rangle + p_3 \tag{A.2}$$

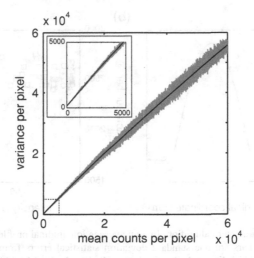

Fig. A.2 PIXIS camera noise curve. The estimation of the photon shot noise can be done by a calibration of the CCD camera noise features. The main figure shows the measured variance of counts versus the mean over the full dynamic range of the camera and averaged over all pixels. The inset details the region typically used in the experiments where the number of counts is limited by the imaging pulse length and its intensity. The slope of the curve extracted by the second order polynomial fit shown in black reveals the photon to count gain factor η necessary for the calculation of the atom number

which is used in the photon shot noise subtraction detailed in Sect. A.3.3. The parameters are $p_1 = -1.63 \times 10^{-6}$, $p_2 = 1.025$ and $p_3 = 130.3$.

Comparison with Gross-Pitaevskii Simulations

The calibration of the c_{ccd} factor is based on the quantum efficiency measurement which itself relies on the power meter calibration. Furthermore the numerical aperture of 0.45 means a coverage of approximately 5% of the solid angle such that a non negligible fraction of the scattered photons enters the objective which is not taken into account in Eq. A.1. Both effects can easily result in a slightly wrong atom number determination.

In order to cross check the inferred number of atoms, we repeatedly image a condensate in a very well known trap configuration using high imaging intensity (typically $50I_{sat}$). In this regime only the linear part of the Beer-Lambert formula contributes (the ratio between the linear and the non-linear part is 2% for the settings used in the experiment.), but the signal to noise ratio is poor. Imaging is done with a very short pulse ($2\mu s$) to avoid any diffusive broadening of the profile due to photon scattering. Since the calculation of the atom number is linear, in situ imaging is possible without worries about the very small size of the condensate (for details see Sect. A.3.2). We calculate the average profile along the long axis of the condensate (much bigger than our optical resolution) for different correction factors c_{gpe} and compare it to theoretical profiles obtained from three dimensional

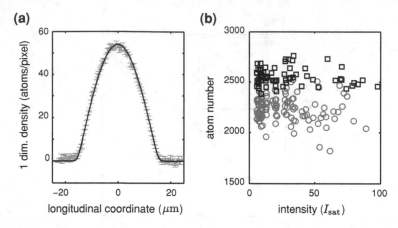

Fig. A.3 Imaging parameter calibration. In **a** an average longitudinal profile of the atomic cloud is shown. The error bars are one standard deviation statistical errors from the averaging over several profiles. The black line is the best fitting prediction obtained from the three dimensional Gross-Pitaevskii equation, where the fitting is done by variation of the total number of atoms in the simulation and the correction factor c_{gpe} when evaluating the data. **b**, Calibration of the Clebsch-Gordan coefficient. Black squares (gray circles) show the deduced atom number for atoms the $F = 2$ ($F = 1$) atoms versus the imaging beam intensity. For the correct Clebsch-Gordan factor no dependence on the imaging intensity is expected

Gross-Pitaevskii equation simulations varying the total numbers of atoms (Fig. A.3.a). We choose the correction factor such that the quadratic deviation between a simulated profile and the measured profile is minimal and find $c_{gpe} = 0.9$, reasonable close to unity.

The Clebsch-Gordan Coefficient

Calibration of the imaging is completed by the measurement of the Clebsch-Gordan coefficient c_{cg}. This is done by imaging a cloud of ultracold atoms with changing imaging intensities I/I_{sat}. A correct Clebsch–Gordan factor means intensity independent atom number measurements. We determine the coefficient c_{cg} by analyzing the obtained pictures assuming different values of c_{cg}. Figure A.3b shows the results of the calibration measurements with a best Clebsch-Gordan coefficient of $c_{cg} = 0.19$ ($c_{cg} = 0.28$) for the $F = 2$ ($F = 1$) atoms.

Knowing all parameters, Eq. A.1 is used to calculate the number of atoms per pixel, after the mean light intensity on the reference picture is normalized to the mean light intensity on the picture containing the atomic signal.[7] The total atom number follows from summation over the region where the atoms are detected. The size of this region is typically chosen to three standard deviations as obtained

---

[7] For the calculation of the mean light intensity the area containing the atomic signal is not taken into account.

from a gaussian fit to the atomic cloud. We checked that neither the detected atom number nor its fluctuations depend critically on the size of the integration area.

A.3.2 Non-linear Effects

Quantitative measurements of the atom number using absorption imaging is limited to atomic clouds with a size larger than the optical resolution of the imaging system. For too small clouds the non-linear part of Eq. A.1 causes an systematic error in the detected atom number. This can be easily understood in the limit where the optical resolution is much larger than the pixel size which is limiting in this case. The detected column density per pixel n is calculated from photon counts integrated over the size of the pixel

$$n \propto \ln \left(\frac{\int_{\text{pix}} N_{\gamma,\text{ref}}}{\int_{\text{pix}} N_{\gamma,\text{pic}}} \right) \tag{A.3}$$

where we omitted the linear term. Correct atom number calculation n_{true} however requires to calculate

$$n_{\text{true}} \propto \int_{\text{pix}} \ln \left(\frac{N_{\gamma,\text{ref}}}{N_{\gamma,\text{pic}}} \right) \tag{A.4}$$

In general $n \neq n_{\text{true}}$ holds such that the calculated atom number is wrong. In our case the pixel size is smaller than the point spread function f of the imaging system. Here the same argument holds, but the main averaging effect is due to the convolution with the point spread function, e.g. the replacement $N_\gamma \rightarrow (N_\gamma * f)$ has to be made, where $(g * f)$ means the convolution of the functions f and g. Figure A.4 shows the measured underestimation of the atom number and the undesired nonlinearity in the atom number calculation for high optical densities and small atomic clouds. We make sure to expand the atomic clouds before imaging to avoid this effect (see Sect. A.2).

A.3.3 Photon Shot Noise Estimation

Our experiments aim to measure atomic fluctuations between two modes, e.g. two neighboring wells of an optical lattice or two internal hyperfine states of the atoms. The atoms are detected via their resonant interaction with the probe light whose noise characteristics add to the atomic noise of interest. In order to minimize this extra noise we take special care to assure photon shot noise limited detection, in particular the absorption pictures have to be free of interference fringes. These fringes originate from motion of the imaging systems optics or air motion in the

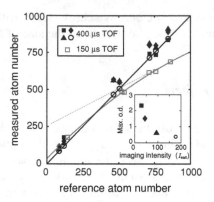

Fig. A.4 Nonlinearity of the absorption imaging system for small atomic clouds. We compare the atom number deduced for different imaging intensities (different shapes of the symbols) and for two different cloud sizes. The cloud size is set by the expansion time prior to the imaging process. For small clouds (gray symbols) and up to intermediate imaging intensities (data taken with $25I_{sat}$) the atom number is systematically underestimated with rising optical density. Linearizing the result around 750 atoms we discover a factor of two smaller slope of the deduced atom number while the absolute number is underestimated by only 15%. This results in a strong bias of the detected atom number fluctuations in this regime. For our squeezing measurements we choose a proper expansion time longer than 400 μs and an intensity of $10-15I_{sat}$ such that the atom number deduction is linear. The reference atom number in this figure was measured for 400 μs expansion time and with high imaging intensity I/I_{sat} (black circles)

light path and lead to an increased background noise level in the images. Our stable imaging system allows for interference fringe free—detection light shot noise limited—pictures and the light shot noise contribution can be inferred pixel by pixel using the camera calibration curve shown in Fig. A.2. We calculate the photonic noise ΔN_γ^2 expected from the measured counts per pixel for both, the picture and the reference picture, using the fit result from Eq. A.2. Standard error propagation of Eq. A.1 allows for a conversion of the photonic variance $\Delta N_{\gamma,\text{pic(ref)}}^2$ into atomic variance:

$$\delta n_{\text{psn,i,j}}^2 = \frac{d^4}{\sigma_0^2} \left\{ \frac{1}{c_{cg}^2} \left(\frac{\Delta N_{\gamma,\text{pic}}^2}{N_{\gamma,\text{pic}}^2} + \frac{\Delta N_{\gamma,\text{ref}}^2}{N_{\gamma,\text{ref}}^2} \right) \right.$$
$$+ \left(\frac{c_{ccd}c_{gpe}}{\tau} \right)^2 \left(\Delta N_{\gamma,\text{pic}}^2 + \Delta N_{\gamma,\text{ref}}^2 \right)$$
$$\left. + \frac{2c_{ccd}c_{gpe}}{c_{cg}\tau} \left(\frac{\Delta N_{\gamma,\text{pic}}^2}{N_{\gamma,\text{pic}}} + \frac{\Delta N_{\gamma,\text{ref}}^2}{N_{\gamma,\text{ref}}} \right) \right\} \qquad (A.5)$$

Here we calculate the photon shot noise contribution $\delta n_{\text{psn,i,j}}^2$ by a expansion up to second order, assuming a gaussian distribution for the photon statistics $p(N_\gamma)$.

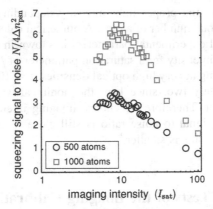

Fig. A.5 Optimum signal to noise ratio. A calculation of the detection signal to noise ratio for a simulated atomic density distribution with 500 and 1,000 atoms reveals the optimal imaging intensity $I/I_{sat} \approx 10$. The vertical axis is scaled to the shot noise limit for the atom number difference $N/4_{psn}$ revealing that—at atomic shot noise level—at least 30% of the total detected noise originates from the imaging process assuming 500 atoms in total

This approximation is justified since we make sure to have more than 200 counts per pixel on all pictures. Photon shot noise on different pixels (i,j) is uncorrelated such that the total photon shot noise contribution in a given area of the CCD chip δn^2_{psn} can be obtained by summation over the variance per pixel.

The total detected atom fluctuations Δn^2_{det} between the modes a or b are the sum of the atomic variance of the quantum state Δn^2, the photon shot noise Δn^2_{psn} and extra technical noise due to experimental instabilities Δn^2_{exp}, e.g. position or magnetic field fluctuations.

$$\Delta n^2_{det} = \Delta n^2 + \Delta n^2_{psn} + \Delta n^2_{exp} \tag{A.6}$$

All these contributions are independent from each other, such that they can be subtracted in order to get the best estimate of Δn^2 if their value is accurately known. The uncertainty in the estimated amount of photon shot noise is given by the accuracy of the camera calibration curve shown in Fig. A.2, where the fit gives an error of 4%.

A.3.4 Signal to Noise Optimization

In the strong saturation regime the imaging light intensity I controls the transparency of the atomic cloud [4]. The imaging signal to noise ratio can be optimized by choosing the proper light intensity depending on the size of the atomic cloud and the total number of atoms. A numerical calculation for the PIXIS camera and for typical experimental parameters is shown in Fig. A.5 and reveals the optimum imaging intensity for a saturation parameter of $I \approx 10 I_{sat}$. One has to be careful when working at very high optical densities, $I \approx 10 I_{sat}$ means an optical density of approximately two, since here the nonlinearity problem detailed in Sect. A.3.2 is strongest. Therefore we choose imaging intensities between $10 I_{sat}$ and $15 I_{sat}$, where the signal to noise ratio is still close to the optimum, but the sensitivity to the cloud size is smaller.[8]

A.4 Independent Tests of the Imaging Calibration

The External Squeezing Experiment

From the discussion above it is clear that correct calibration of the imaging system is not trivial and the results presented in this thesis depend critically on the linearity of the atom number detection. In the first (external squeezing) experiment we used atom number loss in order to check the imaging calibration. We prepared independent Bose–Einstein condensates in the different lattice sites by direct evaporation into a very high lattice situation. The relative atom number fluctuation between different sites is expected to be at the shot noise level in this case. In order to tune the total number of atoms we allow for some loss of atoms and monitor the fluctuations. As shown in Fig. A.6 we find a linear dependence of the measured fluctuations versus the mean total atom number where the slope is compatible with unity within the statistical uncertainties. The dashed line shows the behavior expected for poissonian fluctuations and a correct calibration of the imaging system. Most data points fall within the shown $\pm 20\%$ uncertainty region which we take therefore as the upper bound for possible systematic errors in the imaging calibration for these experiments. The scattering of the data is due to limited statistics (100 measurements per data point). For a more quantitative test we start with a slightly number squeezed state $-3 \, \mathrm{dB} < \xi_N^2 < 0 \, \mathrm{dB}$ obtained after a controlled but fast (20 ms) lattice ramp up to a situation with negligible tunneling. We hold the system in the trap for 10 s after which two-thirds of the atoms are lost. Knowing the loss rates we predict the the expected number squeezing to

---

[8] We expand the cloud prior to imaging, but detection with higher imaging intensity secures even more, that our data is taken in the regime where the imaging calibration is reliable.

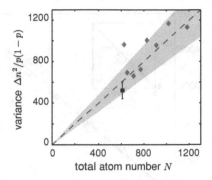

Fig. A.6 Calibration test for the external squeezing experiments. In order to test the imaging calibration we monitor the evolution of number squeezing with particle loss. Loss changes the total number of atoms in the trap and we expect only slight dynamics in the number squeezing for the measured loss rates. The poissonian variance is plotted as the dashed line for reference and the measurements are consistent with the gray uncertainty band of 20%. Data plotted as gray diamonds is measured by condensation into individual lattice sites after different hold times. The black data point correspond to a high statistics measurement ($\approx$ 1,000 measurements) in order to test the calibration with smaller statistical uncertainty. Here we start from a slightly squeezed situation and measure the relative number fluctuations after two-thirds of the atoms are lost. In Sect. 3.4.4 we explain that we expect this data point to lie slightly below the dashed line as found in the measurement

-1.2 dB $< \xi_N^2 < -1$ dB at the time of imaging (see App. B). We measure $\xi_N^2 = -0.7_{-0.7}^{+0.7}$dB where the errors are two standard deviations and the fluctuations where extracted from approximately 1,000 experimental realizations. This result confirms our atom number calibration for this experiment.

The Interferometry and Internal Squeezing Experiment

The squeezing and interferometry experiment based on two internal states of ^{87}Rubidium requires a different calibration, since three pictures, one for each hyperfine state and a reference picture, are necessary to extract the relative atom number fluctuations. The initial quantum state—a maximal Dicke state with all atoms in mode $|a\rangle$—is much better known here as compared to the external case where temperature affects the fluctuations. A coherent spin state centered on the equator of the Bloch sphere with known—shot noise—fluctuations in $\Delta \hat{J}_z^2 = \Delta n^2$ is prepared by a fast $\pi/2$ pulse. The mean total atom number can be experimentally varied by the evaporation ramp without affecting the effective parameters of the system such as tunneling rate or temperature. Therefore it is straightforward to obtain an independent experimental test of the imaging calibration by measuring the relative occupation number fluctuations of the coherent spin state versus the total atom number. Data shown in Fig. A.7 confirm the linear dependence expected where a quadratic fit reveals a slope of 1.01 ± 0.03 (two standard deviation errors) and a small quadratic contribution of $2 \cdot 10^{-5}$.

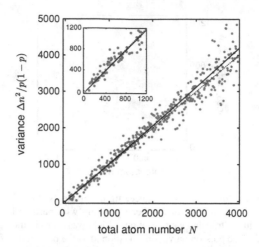

Fig. A.7 Calibration test for the interferometry and internal squeezing experiments. The variance of a coherent spin state on the equator of the Bloch sphere is measured versus the total atom number to test the imaging calibration. In the main figure the total atom number range is extended by the binning technique detailed in Sect. 4.5.1. However this procedure uses information from different sites of the optical lattice rather than changing the population of each individual coherent spin state. Therefore we show the non-binned result in the inset, where only the information from individual wells is used. The measured data confirms the independently obtained calibration of the imaging system

All different combinations of lattice sites were evaluated equivalent to the procedure described in Sect. 4.5.1 in order to expand the range of total atom numbers. The inset shows the same data but without the binning technique such that here only the occupation numbers of the two hyperfine state in a single well contribute. A linear fit reveals a slope of 0.98 ± 0.06 (two standard deviation errors) consistent with the result from the binned data and with a slope of unity as expected for a correct calibration.

Appendix B
Particle Loss and Number Squeezing

For the tests of the imaging calibration in the external squeezing experiments we monitor number squeezing versus particle loss. Therefore it is essential to understand the connection of loss and number fluctuations. The most important processes in this experiment leading to a loss of atoms from the trap are single particle losses with a rate K_1 due to photon scattering or collisions with background gas atoms and three body collisions parametrized by K_3. Spin relaxation loss—a two body process—is negligible for the $|F, m_F\rangle = |2, \pm 2\rangle$ hyperfine states of ^{87}Rubidium [5, 6], however it is the main loss mechanism for the $|F, m_F\rangle = |2, -1\rangle$ state used in the internal squeezing experiments.[9] Here we focus on loss processes relevant for the external squeezing experiment, since in the internal squeezing case the loss happens during the state preparation which requires a more advanced calculation that has been performed in references [8, 9].

One and Three Body Loss and Their Effect on Number Fluctuations

The single particle loss rate is independent of the number of atoms and the loss process results in a random reduction of the atom number in the trap. Therefore it tends to restore binomial fluctuations. However three body loss is a non-linear process requiring the collision of three atoms. The loss rate $L_3 \propto K_3 N^2$ is proportional to the loss coefficient K_3 and the atom number squared resulting in a suppression of atom number fluctuations. A Master equation approach for the joint probability distribution $P(n_l, n_r; t)$ for the number of atoms $n_{l,r}$ in the left and right well can be used to calculate the effect of the particle loss on number squeezing.

---

[9] Close to the Feshbach resonance both two and three body loss become stronger [7] but away from the resonance the lifetime limit is set by the two body process.

C. Groß, *Spin Squeezing and Non-linear Atom Interferometry with Bose–Einstein Condensates*, Springer Theses, DOI: 10.1007/978-3-642-25637-0,
© Springer-Verlag Berlin Heidelberg 2012

$$\frac{\partial P(n_l, n_r; t)}{\partial t} = K_1[(n_l + 1)P(n_l + 1, n_r; t) - n_l P(n_l, n_r; t)]$$
$$+ K_1[(n_r + 1)P(n_l, n_r + 1; t) - n_r P(n_l, n_r; t)]$$
$$+ \frac{K_3}{3}[(n_l + 3)(n_l + 2)(n_l + 1)P(n_l + 3, n_r; t)$$
$$- n_l(n_l - 1)(n_l - 2)P(n_l, n_r; t)]$$
$$+ \frac{K_3}{3}[(n_r + 3)(n_r + 2)(n_r + 1)P(n_l, n_r + 3; t)$$
$$- n_r(n_r - 1)(n_r - 2)P(n_l, n_r; t)] \tag{B.1}$$

here equal loss coefficients for atoms in the left and right well are assumed. Within this description the different moments and correlation functions of $n_{l,r}$ are given by

$$\langle n_l^\alpha n_r^\beta \rangle_t = \sum_{n_l} \sum_{n_r} n_l^\alpha n_r^\beta P(n_l, n_r; t) \tag{B.2}$$

We are interested in the time evolution of the number squeezing parameter $\xi_N^2 = \frac{\langle (n_l - n_r)^2 \rangle}{\langle n_l + n_r \rangle}$, where we assume without loss of generality $\langle n_l - n_r \rangle = 0$:

$$\partial_t \xi_N^2 = \frac{(\partial_t \langle n_a^2 \rangle + \partial_t \langle n_b^2 \rangle - 2\partial_t \langle n_a n_b \rangle)(\langle n_a \rangle + \langle n_b \rangle)}{(\langle n_a \rangle + \langle n_b \rangle)^2}$$
$$- \frac{(\langle n_a^2 \rangle + \langle n_b^2 \rangle - 2\langle n_a n_b \rangle)(\partial_t \langle n_a \rangle + \partial_t \langle n_b \rangle)}{(\langle n_a \rangle + \langle n_b \rangle)^2} \tag{B.3}$$

In order to calculate the time derivatives in this Eqs. B.1 and B.2 are used. In the case of pure one body loss ($K_3 = 0$) the equations can be solved without further approximation and number squeezing evolves like

$$\xi_N^2(t) = 1 - (1 - \xi_{N,0}^2)e^{-K_1 t} \tag{B.4}$$

where $\xi_{N,0}^2$ is the number squeezing at $t = 0$. Note that $e^{-K_1 t} = \frac{N(t)}{N_0}$ is the fraction of total atoms remaining in the trap at time t, meaning number squeezing tends asymptotically to $\xi_N^2 = 0$ dB in the limit where all atoms are lost.

Taking three body loss additionally into account higher moments of the distribution enter the calculation, such that the differential equation can not be written in a closed form any more. A gaussian ansatz has numerically shown to be a good approximation [10] as long as some atoms remain in the trap ($N(t) \gg 1$):

$$P(n_l, n_r) = \frac{1}{\pi \sigma_n \sigma_N} \left(\exp\left(- \left[\frac{n_l + n_r - N}{\sqrt{2}\sigma_N} \right]^2 - \left[\frac{n_l - n_r}{\sqrt{2}\sigma_n} \right]^2 \right) \right) \tag{B.5}$$

with the total atom number $N = \langle n_l \rangle + \langle n_r \rangle$, its variance $\sigma_N^2 = \Delta N$ and the variance of the atom number difference $\sigma_n^2 = \langle (n_l - n_r)^2 \rangle$. Using this ansatz in the master equation we express all higher moments of the gaussian distribution by its

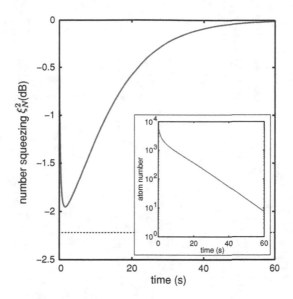

Fig. B.8 Evolution of number squeezing due to one and three body loss. The main graph shows a calculation of the evolution of number squeezing versus the hold time for typical loss parameters of our experiment with atoms in the $|F, m_f\rangle = |2, 2\rangle$ hyperfine state. The initial atom number is 10,000, fluctuations are at shot noise level, the one body loss coefficient is $K_1 = 0.1\,\text{s}^{-1}$ and the effective three body coefficient is $K_3 = 1 \times 10^{-8} s^{-1}$. On intermediate timescales number squeezing develops, but in the long time limit one body loss restores poissonian fluctuations. The dashed line is the limit for pure three body loss. The inset shows the number of atoms remaining in the trap versus time

first two moments (here N, σ_N^2 and σ_n^2) and obtain

$$\partial_t N \approx -K_3 N^3 - K_1 N$$
$$\partial_t \sigma_N^2 \approx K_3(3N^3 - 6N^2\sigma_N^2) + K_1(N - 2\sigma_N^2) \qquad \text{(B.6)}$$
$$\partial_t \sigma_n^2 \approx K_3(3N^3 - 6N^2\sigma_n^2) + K_1(N - 2\sigma_n^2)$$

We kept only the leading order terms which is a good approximation as long as the fluctuations are in the order of shot noise or smaller ($\sigma_N^2 = \mathcal{O}(N), \sigma_n^2 = \mathcal{O}(N)$). Numerical solution of the equations above is straightforward and we obtain the evolution of $\xi_N^2(t) = \frac{\sigma_n(t)^2}{N(t)}$. As shown in Fig. B.8 three body loss can lead to number squeezing on intermediate timescales due to its non-linear dependence on the atom number. The optimum number squeezing achievable is $\xi_N^2 \approx -2.2\text{dB}$ but for our experimental parameters one body loss dominates in the long time limit and restores poissonian fluctuations.

Appendix C
Active Stabilization of Magnetic Fields
Below the Milligauss Level

The performance of the non-linear beamsplitter implemented for the $|1, 1\rangle$ and $|2, -1\rangle$ hyperfine states of ^{87} Rubidium depends critically on the knowledge of the longitudinal position φ of the quantum state at the time of the final rotation pulse (see Sect. 4.5). Fluctuations $\Delta \varphi$ of this angle translate into noise in the population imbalance measured at the output of the non-linear interferometer thus it degrades the overall performance. The most critical external parameter that influences φ is the magnetic field B since it controls the differential energy splitting between the two hyperfine states via the Zeeman effect. Around $B = 9.1$G the detuning caused by magnetic field fluctuations ΔB is

$$\Delta \omega_0 = \Delta B \cdot 2\pi \cdot 10 \text{Hz/mG} \tag{C.1}$$

Despite the usage of a spin echo pulse the system is still sensitive to low frequency phase noise with spectral frequencies between approximately 10 and 300 Hz as shown in Fig. 4.14. In this frequency range we estimate a maximal tolerable magnetic field noise in the order of a few hundred microgauss such that the fluctuations do not dominate the experimental signal.

Low Frequency Field Fluctuations

The spin echo pulse cancels the effect of low frequency magnetic field fluctuations (DC fluctuations) on the acquired relative phase during free evolution within the non-linear beam splitter. However DC fluctuations cause 'shot to shot' errors of the coupling pulses. The sensitivity of the coupling pulses to uncontrolled magnetic field offsets depends on the longitudinal angle $|\varphi_{\text{cpl}}|$ between rotation axis of the pulse and the quantum state. There are two limiting cases: For $|\varphi_{\text{cpl}}| = \pi/2$ a detuning $\Delta \omega_0$ results in an effective Rabi frequency

$$\Omega_{\text{eff}} = \sqrt{\Omega^2 + \Delta \omega_0^2} \approx \Omega(1 + \frac{\Delta \omega_0^2}{2\Omega^2}) \tag{C.2}$$

C. Groß, *Spin Squeezing and Non-linear Atom Interferometry with Bose–Einstein Condensates*, Springer Theses, DOI: 10.1007/978-3-642-25637-0, © Springer-Verlag Berlin Heidelberg 2012

showing only a quadratic dependence in $\frac{\Delta\omega_0}{\Omega}$. When the rotation axis points through the center of the quantum state ($|\varphi_{cpl}| = 0$) the situation is more critical. Detuning causes a change of the rotation axis in polar direction

$$\Delta\theta = \arctan\left(\frac{\Delta\omega_0}{\Omega}\right) \approx \frac{\Delta\omega_0}{\Omega} \tag{C.3}$$

which is linearly dependent on the detuning. These fluctuations contribute most to noise in the occupation number difference if the total pulse angle is $\alpha = \pi$. Therefore we choose the axis of the spin echo pulse to $|\varphi_{cpl}| \approx \pi/2$ for best noise suppression.

Magnetic Field Stabilization

The major problem is caused by magnetic field noise at a spectral frequency of 50 Hz due to the electrical power line. We measure this noise component to be in the order of 4 mG, approximately a factor of ten larger than acceptable. Measurements of the phase stability of the power line signal reveal a coherence time in the order of a few hundred milliseconds. We minimize the degrading effect of this 50 Hz noise by synchronization of the timing of the whole experiment to the power line signal 50 ms before the interferometric sequence. However, we found that further magnetic field stabilization is necessary in our experiment in order to overcome the technical noise problems. We implemented an active feedback loop where we measure the magnetic field as close as possible to the atomic cloud using a fluxgate magnetometer.[10] Since the distance to the atomic cloud is still in the order of 10 cm, the magnetic offset field needs to be homogeneous in a large volume including the magnetic field sensor and the atoms. We installed a pair of quadratic offset coils spaced by 1 m with a side length of 96 cm. Each of these coils has 11 windings and a current of approximately 150 A is necessary to generate a magnetic field of 9 G in the center of the pair. The current is provided by a Delta SM 15-200-D-P104-P145 power supply. We use the internal feedback loop of the power supply in constant voltage mode for coarse control of the magnetic field. The control voltage is generated by a home build fixed voltage source with a relative stability better than 10^{-5} per day. For the Fluxgate based feedback loop we use an extra coil pair winded on top of the first coils. A home build current source allows to change the magnetic field by ca. 200 mG, enough to tune the magnetic field around the Feshbach resonance. We choose the cut off frequency of the feedback loop to a few 100 Hz such that low frequency fluctuations can be compensated to an amplitude in the $100\mu G$ range. Figure C.9 shows the measured noise spectrum with closed feedback loop. In order to avoid local magnetic fields seen by the atoms but not by the Fluxgate sensor we disconnect all other coils close to the experimental chamber by relays.

---

[10] Bartington Instruments, Mag-03MS1000

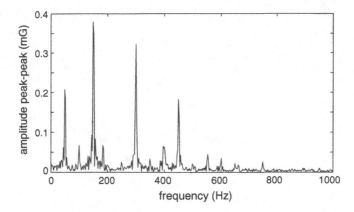

Fig. C.9 Measured magnetic noise spectrum in closed loop configuration. The figure shows the magnetic field noise spectrum obtained from the error signal of the fluxgate sensor feedback loop. The background noise level is low, in the order of $50\mu G$. The main features are the spikes most pronounced at odd multiples of the 50 Hz power line frequency. None of the peaks exceeds the $400\mu G$ level showing the suppression due to the feedback loop, since in unlocked condition the power line noise signal is approximately 4 mG. It is important to note that the measurement is not meaningful in the DC limit since the loop reacts to tilts of the sensor itself or offset voltage drifts, resulting in a change of the magnetic field at the position of the atomic cloud

References

1. Fölling J (2005) Bose–Einstein Josephson tunneling and generation of arbitrary optical potentials. Diploma thesis, University of Heidelberg
2. Ottenstein TB (2006) A new objective for high resolution imaging of Bose–Einstein condensates. Diploma thesis, University of Heidelberg
3. Ketterle W, Durfee D, Stamper-Kurn D (1999) Making, probing and understanding Bose–Einstein condensates. arXiv:cond-mat/9904034v2
4. Reinaudi G, Lahaye T, Wang Z Guéry-Odelin D (2007) Strong saturation absorption imaging of dense clouds of ultracold atoms. Opt. Lett. 32:3143–3145
5. Burt E et al (1997) Coherence, correlations, and collisions: what one learns about Bose–Einstein condensates from their decay. Phys Rev Lett 79:337–340
6. Söding J et al (1999) Three-body decay of a rubidium Bose–Einstein condensate. Appl Phys B 69:257–261
7. Chin C, Grimm R, Julienne P, Tiesinga E (2010) Feshbach resonances in ultracold gases. Rev Mod Phys 82:1225
8. Li Y, Treutlein P, Reichel J, Sinatra A (2009) Spin squeezing in a bimodal condensate: spatial dynamics and particle losses. Eur Phys J B 68:365–381
9. Li Y, Castin Y, Sinatra A (2008) Optimum spin squeezing in Bose–Einstein condensates with particle losses. Phys Rev Lett 100:210401
10. Appmeier J (2007) Bose–Einstein condensates in a double well potential: a route to quantum interferometry. Diploma thesis, University of Heidelberg